WAP: A Beginner's Guide

DALE **BULBROOK**

Osborne/**McGraw-Hill**
New York Chicago San Francisco
Lisbon London Madrid Mexico City Milan
New Delhi San Juan Seoul Singapore Sydney Toronto

Osborne/**McGraw-Hill**
2600 Tenth Street
Berkeley, California 94710
U.S.A.

To arrange bulk purchase discounts for sales promotions, premiums, or fund-raisers, please contact Osborne/**McGraw-Hill** at the above address. For information on translations or book distributors outside the U.S.A., please see the International Contact Information page immediately following the index of this book.

WAP: A Beginner's Guide

Copyright © 2001 by The McGraw-Hill Companies. All rights reserved. Printed in the United States of America. Except as permitted under the Copyright Act of 1976, no part of this publication may be reproduced or distributed in any form or by any means, or stored in a database or retrieval system, without the prior written permission of the publisher, with the exception that the program listings may be entered, stored, and executed in a computer system, but they may not be reproduced for publication.

1234567890 CUS CUS 01987654321

ISBN 0-07-212956-5

Publisher
  Brandon A. Nordin
Vice President & Associate Publisher
  Scott Rogers
Acquisitions Editor
  Michael Sprague
Project Editor
  Monika Faltiss
Acquisitions Coordinator
  Paulina Pobocha
Technical Editor
  Ray Valdez
Copy Editor
  Judith Brown
  Andy Carroll

Proofreader
  John Gildersleeve
Indexer
  Irv Hershman
Computer Designers
  Lauren McCarthy
  Melinda Moore Lytle
  Kelly Stanton-Scott
Illustrators
  Michael Mueller
  Alex Putney
Series Design
  Peter F. Hancik
Cover Series Design
  Amparo Del Rio

This book was composed with Corel VENTURA™ Publisher.

Information has been obtained by Osborne/**McGraw-Hill** from sources believed to be reliable. However, because of the possibility of human or mechanical error by our sources, Osborne/**McGraw-Hill**, or others, Osborne/**McGraw-Hill** does not guarantee the accuracy, adequacy, or completeness of any information and is not responsible for any errors or omissions or the results obtained from use of such information.

ABOUT THE AUTHOR

Dale Bulbrook has been working with computers for over 30 years. Starting as a computer operator on the earliest mainframe computers in 1969, then being promoted to Systems Analyst, and then to Systems Analyst/Programmer. He moved onto PC's and PC programming in 1982. Becoming an accomplished DBase and then Clipper programmer, he has written programs in languages as diverse as Basic and Assembler, Fortran and Visual C++.

He runs his own company, WebDesigns Ltd, and lives in East Grinstead, West Sussex, England—an hour's journey south of London and ten minutes from the Ashdown Forest. He has a beautiful wife named Fareeda; two sons, Tahir and Shaun, and a big yellow Labrador dog called Prince.

This book is dedicated to my mother, Bessie Bulbrook,
who taught me independence and the difference between right and wrong,
and to the memory of my father, James Meredith Bulbrook,
from whom I learned responsibility and duty.
I am privileged to have had you as my parents.

CONTENTS

Acknowledgments	xiii
Introduction	xv

▼ 1 Introducing WAP — 1

What Is WAP?	2
Why is WAP Important?	3
WAP's Benefits for Consumers	5
A History of WAP	6
A Time Before WAP	6
The WAP Forum	7
The Idea of WAP	7
WAP Architecture	8
The WAP Model	9
Wireless Markup Language (WML)	10
Evolution of WAP	10
Adapting to the Restrictions of the Wireless Network	10
The Business Case for WAP	15
WAP Services	15
Why WAP?	20
The Future of WAP	21

2 What Makes a Good WAP Application? ... 23
- The User's Point of View ... 24
  - Ease of Use ... 25
  - Designing for Users ... 25
- What Are the WAP Micro-Browser Issues Today? ... 27
  - Writing a Generic WML Interface ... 28
  - Targeting Your Market Micro-Browser ... 29
- How to Design a Good WAP Application ... 30
  - The Application-Design Process ... 32
  - Common Design Mistakes ... 37

3 The User Interface ... 39
- User Interface Basics ... 40
- Low Bandwidth ... 42
- Small Screen Size ... 42
- Text Entry ... 43
  - Number of Keystrokes ... 43
  - Password Text Entry ... 44
  - Application Personalization ... 44
  - Data Field Entry ... 45
- Using the Cache ... 45
- Types of WML Cards ... 47
  - Choice Cards ... 47
  - Entry Cards ... 49
  - Display Cards ... 49
- The "Back" Button ... 50
- Graphics ... 51

4 WAP Development Tools and Software ... 53
- Editors and Emulators ... 55
  - WAP Editors ... 55
  - WAP Emulators ... 58
- Software Developer Kits (SDKs) and Integrated Development Environments (IDEs) ... 60
- Converting Images ... 62
  - Specification of Well-Defined WBMP Types ... 62
- Summary ... 63

5 Working with WML ... 65
- WML Basics ... 66
  - WAP and the Web ... 67
  - Writing WML Code ... 68

The "Hello World" Example	70
The Document Prologue	70
The Deck Header	72
The First Card	72
The Second Card	72
The Deck Footer	73
A Services Site Example	73
Using Multiple Decks	73
Building the Services Site	74
What's in a Card	75
Graphics	76
The Services Site with Graphics	78
Creating Links	80
The WML Site with Links	81
Templates	83

▼ 6 Interactivity: Forms and User Input — 85

The Options Menu (Select)	86
Selection on the Nokia	87
Selection on Phone.com	88
Option Groups	88
Templates Revisited	92
The Do Element	94
Events	98
Onenterbackward	98
Onenterforward	101
Onpick	102
Ontimer	103
Variables	104
Using Variables	105
Other Ways of Setting Variables	108
The Input Tag	110
Data Formatting	112
Summary	116

▼ 7 Adding Functionality with WMLScript — 117

What is WMLScript?	118
The Rules of WMLScript	119
Case Sensitivity	120
Whitespace and Line Breaks	120
Comments	120
Statements	122
Code Blocks	122

Variables	123
Variable Scope	124
Operators	124
Assignment Operator	125
Arithmetic Operators	125
Bitwise Operators	126
Increment and Decrement Operators	126
Logical Operators	128
Comparison Operators	129
String Concatenation	131
The Comma	131
The typeof Operator	132
The isvalid Operator	133
The Conditional Operator	133
Operator Precedence	134
Control Constructs	135
If Statements	135
While Statements	137
For Statements	137
Stopping Loops or Skipping Unnecessary Loop Statements	138
Reserved Words	140
Functions	141
Parameters	141
Calling Functions	142
The Standard Libraries	145
The Dialogs Library	146
The Float Library	146
The Lang Library	147
The String Library	149
The URL Library	150
The WMLBrowser Library	152
Arrays	152
Pragmas	154
External Files	155
Access Control	156
Metadata	157
General Coding Principles	157

▼ 8   Database-Driven WAP   159

Active Server Pages	161
ASP and WAP	162
The ASP Object Model	164

ActiveX Data Objects (ADO) 168
    Physically Connecting to the Database 168
    Querying the Database 169
    Using the Returned Data 169
    Tidying Up 170
    Some Additional Notes about Connections 170

▼ 9   A Dynamic WAP Application 171

Worldwide-Dance-Web for WAP 172
    Data Flow 172
    Building the Database 173
    Writing the Code 176
Summary 193

▼ 10   Converting Existing Web Sites 195

Why Convert an Existing HTML Web Site to WAP? 196
    What Should You Convert? 196
    Methods of Conversion 197
A Demonstration HTML Conversion 202
Summary 210

▼ 11   M-Commerce and Security 213

Types of Security and Why It Is Necessary 214
    What Is an Acceptable Level of Security? 215
    How Secure Is WAP? 215
A Brief History of Encryption 216
    Cryptography 216
Wireless Transport Layer Security 219
    The Handshake 220
Summary 221

▼ 12   Push Technology and Telematics 223

Push Technology 224
    The Push Framework 224
Telematics 228
    Location-Sensitive Information 228
    Applications for Telematics 229
Push and Telematics Together 230
    User Privacy 231
Summary 231

13 What the Future Holds ... 233
Technology with Users in Mind ... 234
Bluetooth — Cutting the Cords ... 235
VoiceXML—a New Slant on "Walkie/Talkie" ... 237
Telematics—We Know Where You Are ... 238
Bringing It All Together ... 239

14 WMLScript Reference ... 241
Case Sensitivity ... 242
Whitespace and Line Breaks ... 242
Comments ... 243
Constants ... 244
    Integer Constants ... 244
    Floating-point Constants ... 244
    String Constants ... 245
    Boolean Variables ... 246
    Invalid Variables ... 246
Reserved Words ... 247
Variables ... 248
    Variable Declaration ... 248
    Variable Scope and Lifetime ... 248
Data Types ... 249
Pragmas ... 250
    External Files ... 250
    Access Control ... 251
    Metadata ... 252
Operators ... 253
    Assignment Operators ... 253
    Arithmetic Operators ... 254
    Logical Operators ... 255
    String Operators ... 256
    Comparison Operators ... 256
    Comma Operator ... 257
    Conditional Operator ... 258
    typeof Operator ... 258
    isvalid Operator ... 259
Expressions ... 259
Functions ... 260
    Function Declarations ... 260
    Function Calls ... 260

Statements	262
Empty Statements	262
Expression Statements	262
Block Statements	262
Variable Statements	263
If Statements	263
While Statements	264
For Statements	264
Break Statements	265
Continue Statements	265
Return Statements	266
Libraries	266
Notational Conventions	266
Lang Library	267
Float Library	272
String Library	275
URL Library	284
WMLBrowser Library	290
Dialogs Library	293
Console Library	295
▼ Glossary	297
▼ Index	307

ACKNOWLEDGMENTS

I have seen many acknowledgements in books where authors say how wonderful and supportive their spouse and family were "while writing this book." Until now, my first book, I just skimmed over these remarks with the thought, "Yeah well, you have to say that, I guess." It was only when I got involved in this book and actually sat down to write it, that I discovered it is impossible to take on a project of any size like this unless you really do have the full support of your family. Being chained to a desk and computer for months, every evening and every weekend, requires a fantastic amount of commitment from the rest of your family. And as I have been (most of the time) having fun, it is they who have had to do the suffering. With this in mind, I want to thank and acknowledge my wife, Fareeda, and my two sons, Tahir and Shaun, for putting up with me while I have been writing this book and for supporting me as fully as they have done.

I also want to thank the team at Osborne/McGraw-Hill who have been far more patient with me than they had any reason to be, and for the assistance of the staff there. In particular, Monika Faltiss, Paulina Pobocha, the copy-editors: Judith Brown and Andy Carroll, for correcting all of my mistakes and bad grammar, and Michael Sprague for being so patient.

Grateful acknowledgments are also due to Phone.com (now OpenWave), and Nokia for their development software and illuminating documentation, and to the WAP Forum (http://www.wapforum.org) for organizing and codifying the WAP, WML, and WMLScript specifications so that the rest of us could refer to it.

I also thank my friends and associates who had to put up with my intense preoccupation and mood swings, especially Tim Roser to whom I feel I was particularly mean.

And finally, my thanks to Steven Lee, without whom this project would never have gotten started.

INTRODUCTION

I have written this book specifically for beginners. I make no apologies to those who are already proficient in WAP to some degree, or who are already programming in another scripting language. After more than 30 years working with and around computers, I know that there is always something else to know or find out, even in those subjects that I think I know well.

So I have tried hard to make this book understandable to anybody who wants to write real WML sites for WAP devices, or for anybody who just wants to know what "this WAP thing" is all about.

I have also found that being interested in different fields can provide unlooked for solutions to problems that come up every day. Who would have thought that knowing WML for devices with small screens would make me reexamine the way I write HTML for Internet web sites every day? Who would have thought ten years ago that knowing how to code in C would make picking up WMLScript so easy? Or that getting a basic understanding of XML and how it works would make WML such a breeze to work with?

All of the things that we do and experience can have a direct impact on everything else that we do in life. (If you want to learn how to be more patient with others, go and write a word processor in COBOL. If you want a *truly* mind altering experience, try writing and working with an application that uses dynamic multi-dimensional arrays.)

If I have done my job well, you will be able to create some good applications while reading this book and some great applications by the end of the book. If I have done my job very well, then you will be able to apply some of the data in this book to other related fields and make those areas better than they were already. The section on how to make an application more usable to the end user, for example, applies to any application on any machine where you are having to deal with user interaction, whether it is a mobile device or not.

Above all, you should always bear in mind that WAP, and in particular WML and WMLScript, are just tools that can be harnessed to make your imagination a reality. I have tried to paint some pictures of what is possible, and to show you the rudiments of how it can be accomplished.

I have tried to make the learning curve as painless as possible. I know what it is like to get thrown in at the deep end of the pool, and struggling to make sense of anything at all. I also know what it's like to be treated like a little kid, ("This big box here is called a computer. That's a big word, isn't it?"), and so I have tried to avoid being patronizing. You are obviously literate, intelligent, and interested; otherwise you wouldn't even be reading this introduction in the first place. I promise to treat you as such throughout the rest of this book.

Here's to your "killer app"!
Dale Bulbrook

CHAPTER 1

Introducing WAP

It is easy to imagine doing business with your customer anytime, anywhere in the world. It's a daily reality on the Internet. But imagine reaching customers who would never think to use a computer, or who are simply too busy to use one.

WAP is an abbreviation for Wireless Application Protocol, and very simply put, it is what makes it possible to access the Internet via wireless devices such as mobile phones and personal digital assistants (PDAs). The goal of this book is to give you the background you need to write effective, simple WAP applications that will run well and be useful to their users.

There are millions of mobile phones all over the world, and they are being used by everyone from executives in New York to taxi drivers in Istanbul. These are millions of consumers in thousands of cities around the world. With WAP, each of those phones can be used for comparing prices, selecting products, purchasing, and tracking orders.

Mobile commerce has been called the next big growth opportunity. It combines the two most explosive technologies of the new economy—the Web and wireless communications. However, no matter how powerful the Web becomes, as long as it is desk-bound on PCs, it will restrict people by time and location. Wireless devices are paving the way for people to interact, inform, and communicate on the move. Today there are already over 200 million wireless subscribers, and by 2003 it is predicted that there will be more than one billion.

WHAT IS WAP?

WAP stands for Wireless Application Protocol. Per the dictionary definition for each of these words, we have:

- ▼ **Application** A computer program or piece of computer software that is designed to do a specific task.
- ■ **Wireless** Lacking or not requiring a wire or wires: pertaining to radio transmission.
- ▲ **Protocol** A set of technical rules about how information should be transmitted and received using computers.

WAP is the set of rules governing the transmission and reception of data by computer applications on, or via, wireless devices like mobile phones. (I say "like" mobile phones, because the mobile phone is no longer be considered to be just a phone, but a communications device capable of sending and receiving communications in all sorts of different forms.)

As a matter of fact, WAP is not actually one single protocol. It is a collection of protocols and specifications that cover everything from how the WAP device and the user agent should work, to how the transport protocols interact with the bearers themselves.

NOTE: User agent is not just another fancy term for a mobile phone. A *user agent* is any WAP device, whether it is a mobile phone, a handheld device such as the Palm or HP personal digital assistants, a pager, or even a household refrigerator that has been WAP enabled.

WAP is a standardized technology for cross-platform, distributed computing, very similar to the Internet's combination of Hypertext Markup Language (HTML) and Hypertext Transfer Protocol (HTTP), except that it includes one vital feature: optimization for low-display capability, low-memory, and low-bandwidth devices, such as personal digital assistants (PDAs), wireless phones, and pagers.

The major accomplishment of WAP is that it has managed to overcome the drawbacks of handheld devices:

- They have small screens.
- They don't have a lot of free memory to run applications of any size.
- The bandwidth is restricted to 9,600 bits per second.

All of these points are liable to change at any time, and probably sooner rather than later. In the meantime, however, all of these points conspire to make life very difficult for the aspiring WAP developer.

WAP allows wireless devices to view specifically designed pages from the Internet, using only plain text and very simple black-and-white pictures. The WAP programming code at the Web site has to be explicitly designed and written for the micro-browser used in that specific model of WAP device. The pages themselves have to be small, because the data speed on mobile phones is limited—a lot slower than a domestic modem. Also, the WAP-enabled devices have screens of different shapes and sizes, so the same page can look very different depending on the actual device you are using, quite independently of the version of the micro-browser being used in the phone itself.

Why is WAP Important?

There are only a few core industries that will continue to exist and develop regardless of what else may happen to the society that we live in. These are things like food production, food distribution, entertainment, housing and communications. You only have to look at the history books to realize very quickly that the financial institutions, insurance companies, stock markets and many so-called luxury items are historically brand new, and therefore not basic to man's survival.

You can live perfectly well without an insurance policy or a bank account, but you could never live without food, or a place to sleep that is somewhat protected from the elements. These are the basics of survival for meat bodies.

The colossal growth of the Internet can seem pretty staggering, unless one takes into account that it covers the areas of communication and entertainment very well.

The growth in sales of mobile phones across the entire planet could also appear in credible, unless one remembers that man is a social animal, and likes to talk. It was always important to let your mother know where you were after dark, and it always will be important, regardless of the technology that is available to let her know.

Until the first WAP devices emerged, the Internet was the Internet and a mobile phone was a mobile phone. You could just surf the Net, do serious research, or be entertained on the Internet using your computer (see Figure 1-1), and you could talk to your

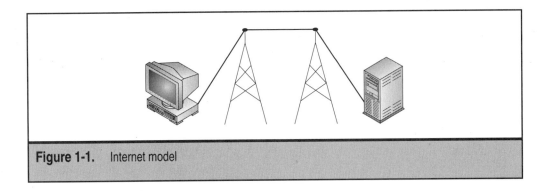

Figure 1-1. Internet model

mother, colleagues, and friends on your mobile phone (see Figure 1-2). They ran on different physical global networks, had completely separate functions, and had no areas of crossover aside from Short Message Service (SMS, see glossary) text messaging, but that was only from mobile phone to mobile phone).

What has happened with the appearance of WAP on the scene is that we have the massive information, communication, and data resources of the Internet becoming more easily available to anyone with a mobile phone or communications device.

Anyone who has tried to access the Internet by using a laptop and a mobile phone knows that the quality of access we expect when using the Internet at the office or at home are not fulfilled. In fact, it is usually an irritating, frustrating and exhausting experience.

Now the playing field has changed significantly. WAP is designed to meet the restrictions of a wireless environment—limitations in both the network and the client have been taken into consideration. As well as being able to talk to your friends and colleagues directly from anywhere in the world with your mobile phone, you can now, with the very same device, get the current prices of your stocks, find out the latest news, and read your e-mail (see Figure 1-3). You can even instruct your service to send you messages telling you the latest scores of your favorite sports team or stock movements, as soon as they happen.

Service providers also benefit from this connectivity, since their services can be deployed independently of the locations of the users. The services are created and stored centrally on a server, and it is very easy to change them according to customer requirements. By using off-the-shelf tools, services can be created with a minimum of effort, providing an extremely short time to market.

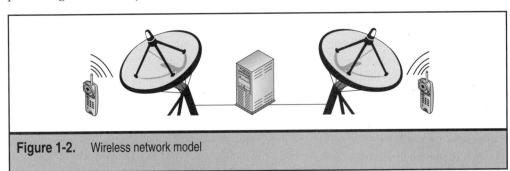

Figure 1-2. Wireless network model

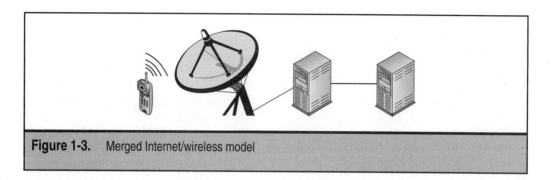

Figure 1-3.   Merged Internet/wireless model

It is this merging of the massive amounts of data available from the Internet with the portability and instantaneous communication of the mobile phone that is the true advantage of WAP.

As more and more Web sites become WAP enabled, more data becomes available instantly through your mobile phone. In fact, the very term mobile phone is misleading, as the phone can now already do so many additional things—the addition of WAP has turned it into a true multifunctional communications device rather than just a phone.

WAP's Benefits for Consumers

It is essential that consumers benefit from using WAP-based services; otherwise there will be no incentive for anybody concerned, and no real reason for WAP to exist.

The key benefits of WAP from a consumer's viewpoint can be summarized as follows:

- It's portable
- It's easy to use
- You have access to a wide variety of services on a competitive market
- Services can be personalized
- You get fast, convenient, and efficient access to services
- WAP devices are available in various forms (pagers, handheld PCs, phones, and so on)

Already in Finland you can use your WAP mobile phone to buy soft drinks from a vending machine, operate jukeboxes, and even purchase car washes.

An Israeli firm has released a software system that not only lets you buy your favorite soft drink using your mobile phone or your PDA and charge it straight to your phone bill, but it simultaneously shows you film trailers from the latest movie releases or the latest sports results. They are constantly expanding the software's applications. The idea is that eventually a parking meter will be able to page you via your phone if you're running overtime, and let you add more money to it remotely.

Service providers will offer news, weather, local restaurant and cinema information, and traffic reports, or some combination thereof. You'll be able to download your horoscope, get real-time traffic directions and updates from a GPS (Global Positioning System) that knows exactly where your car is, and do all of your banking and shopping via your phone. Your voice software will read your e-mail aloud to you in the car and allow you to send voice mail to anyone just as easily.

Your home, too, will go on getting more networked, until there'll be nothing mechanical that you can't control via your mobile device and a smart interface. You'll be able to switch your house lights, heating system, or VCR on and off remotely. One refrigerator manufacturer (Electrolux) is already manufacturing something called the ScreenFridge—if you scan the barcode of an empty carton, it automatically orders a replacement via the Internet.

Eventually you might have a piece of household software that keeps track of what's in the refrigerator. You plan the menu for a dinner party and the software will log itself on to the Internet to order home delivery of whatever you need to bring off a successful meal.

One very important thing to bear in mind is that at the time of writing, WAP is still very much in its infancy. Although the seeds have been germinating since 1990, it is only now really seeing the light of day as the technology is becoming available to deliver it.

A HISTORY OF WAP

WAP could very roughly be described as a set of protocols that has inherited its characteristics and functionality from Internet standards and from standards developed for wireless services by some of the world's leading companies in the business of wireless telecommunications.

A Time Before WAP

In 1995 Ericsson initiated a project whose purpose was to develop a general protocol, or rather a concept, for value-added services on mobile networks. The protocol was named Intelligent Terminal Transfer Protocol (ITTP), and it handled the communication between a service node, where the service application is implemented, and an intelligent mobile telephone. The ambition was to make ITTP a standard for value-added services in mobile networks.

During 1996 and 1997, Unwired Planet, Nokia, and others launched additional concepts in the area of value-added services on mobile networks. Unwired Planet presented the Handheld Device Markup Language (HDML) and Handheld Device Transport Protocol (HDTP). Just as HTML is used on the Web, HDML is used for describing content and user interfaces, but it is optimized for wireless Internet access from handheld devices with small displays and limited input facilities. In the same manner, HDTP can be considered a wireless equivalent of the standard Internet HTTP; it is a lightweight protocol for performing client/server transactions.

In March of 1997, Nokia officially presented the Smart Messaging concept, an Internet-access service technology specially designed for handheld GSM (Global System for Mobile Communications, see glossary) devices.

The communication between the mobile user and the server containing Internet information uses Short Message Service (SMS) and a markup language called Tagged Text Markup Language (TTML). Just like HDML, this language is adapted for wireless communication—for narrowband connections.

With a multitude of concepts, there was a substantial risk that the market would become fragmented, a development that none of the involved companies would benefit from. Therefore, the companies agreed upon bringing forth a joint solution, and WAP was born.

The WAP Forum

On June 26, 1997, Ericsson, Motorola, Nokia, and Unwired Planet (which became Phone.com and has since merged with Software.com to become Openwave Systems Inc.) took the initiative, and in December 1997 the WAP Forum was formally created. The WAP Forum's mission was to bring the convenience of the Internet into the wireless community, and after the release of the WAP 1.0 specifications in April 1998, WAP Forum membership was opened to all.

Its membership roster now includes all of the computer industry heavyweights, such as Microsoft, Oracle, IBM, and Intel, along with a couple of hundred other companies, including carriers, manufacturers, WAP application developers, and so forth. With over 90 percent of the mobile phone handset manufacturers being represented at the WAP Forum, WAP is assured of being the primary way of accessing mobile content on the Internet.

The Idea of WAP

According to the WAP Forum, the goals of WAP are that it:

- ▼ Create a global wireless protocol to work across differing wireless network technologies specification Independent of wireless network standards
- ■ Submit specifications for adoption by appropriate industry and standards bodies
- ■ Enable content and applications to scale across a variety of transport options
- ■ Enable content and applications to scale across a variety of device types
- ▲ Be extensible over time to new networks and transports

By addressing the constraints of a wireless environment, and adapting existing Internet technology to meet these constraints, the WAP Forum has succeeded in developing a standard that scales across a wide range of wireless devices and networks. The standard is license-free, and it brings information and telephony services to wireless devices. To access these services, WAP utilizes the Internet and the Web paradigm. WAP scales across a broad range of wireless networks, implying that it has the potential to become a global standard and that economies of scale can thus be achieved.

Some key features offered by WAP are the following:

- ▼ **A programming model similar to the Internet's**   Reuse of concepts found on the Internet enables a quick introduction of WAP-based services, since both service developers and manufacturers are familiar with these concepts today.
- ■ **Wireless Markup Language (WML)**   This is a markup language used for authoring services, fulfilling the same purpose as HTML does on the Web. In contrast to HTML, WML is designed to fit small handheld devices.
- ■ **WMLScript**   WMLScript can be used to enhance the functionality of a service, just as, for example, JavaScript can be utilized in HTML. It makes it possible to add procedural logic and computational functions to WAP-based services.
- ■ **Wireless Telephony Application Interface (WTAI)**   The WTAI is an application framework for telephony services. WTAI user agents are able to make calls and edit the phone book by calling special WMLScript functions or by accessing special URLs. If one writes WML decks containing names of people and their phone numbers, you may add them to your phone book or call them right away just by clicking the appropriate hyperlink on the screen.
- ▲ **Optimized protocol stack**   The protocols used in WAP are based on well-known Internet protocols, such as HTTP and Transmission Control Protocol (TCP), but they have been optimized to address the constraints of a wireless environment, such as low bandwidth and high latency.

The opportunity of creating wireless services on a global basis will attract operators as well as third-party service providers, resulting in both co-operation and competition that do not exist today. WAP provides a means to create both services that we are familiar with on the Web today and telephony services.

WAP ARCHITECTURE

This book is concerned only with the first layer of the WAP architecture—the Application Layer—which includes Wireless Markup Language (WML) and WMLScript. However, we'll take a brief look at the whole WAP architecture here to get an overall picture of the technology being used and to outline the most important features provided by WAP. If you are interested in any of the remaining layers, you should check out the specification documents at http://www.wapforum.org.

As already stated, WAP is designed in a layered fashion so that it can be extensible, flexible, and scalable. As a result, the WAP protocol stack is divided into five layers:

1. **Application Layer**   Wireless Application Environment (WAE)
2. **Session Layer**   Wireless Session Protocol (WSP)
3. **Transaction Layer**   Wireless Transaction Protocol (WTP)

4. **Security Layer**   Wireless Transport Layer Security (WTLS)
5. **Transport Layer**   Wireless Datagram Protocol (WDP)

Each of these layers provides a well-defined interface to the layer above it. This means that the internal workings of any layer are transparent or invisible to the layers above it.

This layered architecture allows other independent applications and services to utilize the features provided by any of the WAP layers, making it possible to use the WAP layers for services and applications that aren't currently specified by WAP.

Because the WAP protocol stack is designed as a set of layers, that also means that it becomes extendable and future-proof. Any layer can be extended or changed as necessary or desired. As long as the interfaces between the layers are consistent, any individual layer can be changed without affecting the remaining layers in the slightest.

For example, you could amend the security layer to change the encoding algorithm completely without affecting the writing of WML or WMLScript. Conversely, you could change the Wireless Application Environment (WAE) layer to add a whole new set of tags to the Wireless Markup Language (WML) which is part of this layer, but this would still be transmitted in the exact same way by the network to the phone.

The WAP Model

When it comes to actual use, WAP works like this:

1. The user selects an option on their mobile device that has a URL with Wireless Markup language (WML) content assigned to it.
2. The phone sends the URL request via the phone network to a WAP gateway, using the binary encoded WAP protocol.
3. The gateway translates this WAP request into a conventional HTTP request for the specified URL, and sends it on to the Internet.
4. The appropriate Web server picks up the HTTP request.
5. The server processes the request, just as it would any other request. If the URL refers to a static WML file, the server delivers it. If a CGI script is requested, it is processed and the content returned as usual.
6. The Web server adds the HTTP header to the WML content and returns it to the gateway.
7. The WAP gateway compiles the WML into binary form.
8. The gateway then sends the WML response back to the phone.
9. The phone receives the WML via the WAP protocol.
10. The micro-browser processes the WML and displays the content on the screen.

WAP makes use of the Internet model to provide a flexible service platform. However, in order to accommodate wireless access to the information on the Web, WAP has been optimized to meet the restrictions of a wireless environment.

Wireless Markup Language (WML)

Pages or services created using HTML do not work very well on small handheld devices, since they were specifically developed for use on desktop computers with larger color screens. Also, low bandwidth wireless bearers wouldn't be suitable for delivering the large files that HTML pages often consist of. Therefore, a markup language specifically adapted to these restrictions has been developed—WML.

WML provides a navigation model for devices with small display screens and limited input facilities (no mouse and a limited keyboard). In order to save valuable bandwidth in the wireless network, WML can be encoded into a compact binary format for transmission between the phone and the network, and vice-versa. Encoding WML is one of the tasks performed by the WAP gateway, which is the entity that connects the wireless domain with the Internet.

WAP also provides a means for supporting more advanced tasks, comparable to those solved by using JavaScript in HTML. The solution in WAP is called WMLScript.

WML is very similar to the HTML used to write current Web sites. It is simple enough that any developer currently used to HTML can cross-train in a matter of hours. Naturally, there are some differences between HTML and WML, as WML has to be very simple. There are no nested tables, only very basic font control, and the pages (or *decks* as they are called) have to be quite small, so that they do not take ages to download at the current wireless data transfer speed of 9,600 bps.

Evolution of WAP

If you have been around the Web for a while, you may remember the earliest versions of Web browsers. While they were promoted as the next killer app, they were buggy, unreliable, and completely inconsistent in displaying the same code in different versions. (Very much like today's browsers, in fact!)

Unfortunately, WAP has to go through the same developmental sequence. Currently, no two mobile phones or devices are alike, and even the same model of phone can have different versions of the same micro-browser inside it. Developing code for such a situation is nightmarish, just as when Netscape initially introduced layers to their browser. What makes it much harder for WAP developers to gain public acceptance of their applications is that the whole subject of mobile communications and data transmission has been taken over by the marketing departments of the major manufacturers. The public is being incorrectly sold on the idea of "surfing the Web on your mobile telephone," and people are being badly disappointed because the screen of a mobile phone is *never* going to look like the current version of Netscape or Internet Explorer.

However, there is a definite place for WAP technology in the world, and the applications being developed all over the world by some very smart people are going to make a real difference in the way we communicate, do business, and spend our leisure time.

Adapting to the Restrictions of the Wireless Network

Most WAP devices will be mobile phones, but it is important to remember that WAP is not in any way limited to phones. WAP scales across a broad range of wireless networks and bearers, and therefore it is designed to allow access to services via the Internet using Short Message Services (SMS) as well as fast packet data networks, such as General Packet Radio Service (GPRS). WAP can offer services and applications similar to the ones you find on the Internet, but in a very thin client environment, and by thin, I mean that they are limited by several factors: low bandwidth, high latency, limited connection stability, small display size and limited input facilities, memory, CPU, and battery power.

How well a WAP application works is up to the new wave of designers and developers, and while it's true that WAP currently limits the developers in many ways, the technology is new, and newer standards are being evolved all the time. Every professional developer knows that there are ways around almost every obstacle.

The most important limitations in WAP networks are explained in the following sections.

Low Bandwidth

The size of an average HTML page these days, including graphics, is around 20KB. With a 56 Kbps modem, the download time for this page would be in the region of 4 seconds. As the bandwidth of a wireless network is around 9.6 Kbps, however, the download time for the data equivalent of just that one page would be around 17 seconds. That is not making any allowances for the network itself being slow due to congestion, or for latency (which will be covered in a moment). The majority of mobile users are not aware of access speeds, and they should not have to care about the differences in access methods to get the same perception of performance.

WAP addresses this bandwidth issue by minimizing the traffic over the wireless interface. WML and WMLScript are binary encoded into a compact form before they are transmitted, in order to minimize the bandwidth restriction.

The Wireless Session Protocol (WSP) layer, which is WAP's equivalent to HTTP on the Internet, is also binary encoded for the same reasons. In addition, WSP supports both sessions that can be suspended and resumed, and header caching. This all saves valuable bandwidth, since session establishment only needs to be done once in an average session.

The Wireless Transaction Protocol (WTP), which is WAP's equivalent to the Internet's TCP, is not only designed to minimize the amount of data in each transaction, but also the number of transactions.

Using Graphics   You can use pictures, but only in black and white. There are two reasons for this. The first is that most WAP devices currently available have only mono LCD screens, not color. But the main reason is speed of transfer. Pictures take a while to download, and keeping them simple will reduce the amount of time it takes to download them.

As always, though, this depends on the content. Instead of downloading icons or graphics as fluff, as on most Web pages, graphics should be used only where they can be very effective, such as for displaying maps. One of the first applications of WAP was for the Paris Metro—offering maps to WAP-enabled phones. To give you some idea of how good some simple black-and-white maps can be, here are a couple of examples from www.webraska.com, showing where current traffic black spots are in the road systems concerned. See Figure 1-4.

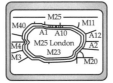 

Figure 1-4. Two digited maps: London's orbital motorway and a section of Paris

This is the kind of application where WAP can really shine. By providing instant visual information that the user can call up and absorb within seconds, it makes the data both extremely useful and usable to the busy user on the move.

High Latency

Latency is the round-trip delay between something being sent on the network, and an acknowledgement obtained that what was sent was received ok. This latency is the time delay introduced by the cumulative effects of software and hardware as a message passes through a network. For example, this is the same principle that can sometimes be noticed on a long distance telephone call when you hear an echo of your voice.

All of the information coming from the Internet and going to the mobile phones has to go through various elements in the mobile network, each one introducing a small delay. Also the wireless interface has a very limited bandwidth, reaching a maximum of 9.6 Kbps, while on a wired network this would be a minimum of 28 Kbps. All the messages to a wireless device therefore have to go through this bottleneck of data transfer, as well as additional software layers and physical network devices, like radio transmitters. When you add in the effects of the standard Internet protocols, which send many large non-optimized messages (because that is the way that the protocols were originally designed), this can result in a very large latency, or round trip delay, for each message.

When a packet of sent data is not acknowledged by the remote entity within a fixed period of time, known as the retransmission timer value, then the TCP layer at the sending end has to resend the packet of data. An *average* latency, or delay, is around half a second in a wireless network, and much less in a wired network simply because there are less software layers to go through, and the available bandwidth is much higher as wireless transmission to a device is not necessary.

The problem is that in some cases, this latency, or delay, on a wireless network can go up to ten seconds or higher. As a result, a TCP layer that works just fine on a normal wired network, when used to carry data across the internet for a wireless device request, might retransmit a lot of data packets because it would believe that the data has been lost in transit. This could cause severe network congestion on a wireless network.

As a result, the application environment in WAP (the WAE) uses the concept of scripting, meaning that round trips between a client and a server can be avoided in certain situations, such as for validating user input. The Wireless Telephony Application

Interface (WTAI) environment addresses the latency problem (and that of low bandwidth) by introducing the repository, which is a persistent storage area used to hold services that should be started in response to an event in the mobile network (such as for an incoming call). Since these services are available immediately, no round trips to the server are needed, and so real-time handling is made possible.

Less Connection Stability and Unpredictable Bearer Availability

Wired network access provides a more or less reliable connection to the network. That is not the case in wireless networks, where the bearers might be inaccessible for shorter or longer periods of time due to fading, lost radio coverage, or deficient capacity. If you have ever lost a connection when you were driving in your car, you will know just how frustrating this can be.

The architects of the WAP protocols infrastructure, when putting together the specifications for WAP, have taken the problem of connection stability into account and have designed into the layers the following features:

- ▼ The sessions supported by the Session layer are assumed to be long-lived, so the problem of lost connections is addressed by allowing lost sessions to be resumed, even when dynamically assigned IP addresses are used.
- ■ The Wireless Transaction Protocol (WTP) layer in WAP has been kept very simple compared to TCP, which is used on the wired Internet. Since no connection is set up, the effects of lost connections and other times of inactivity are minimized.
- ▲ The unpredictable nature of a wireless connection means that small segments of a message are often lost, and WTP supports selective retransmission of data, meaning that only the lost segments are retransmitted and not the entire message as in TCP.

These are very clever solutions to what are otherwise major problems.

Small Display

Instead of using the flat document structure that HTML provides, WML structures its document in *decks* and *cards*. A *card* is a single unit of interaction with the end-user, such as a text screen, a selection list, an input field, or a combination of those. A card is typically small enough to be displayed even on a small screen.

When an application is executed, the user navigates through a series of cards—the series of cards used for making an application is collected in a deck.

Here is an example of a simple deck containing three cards in Wireless Markup Language. This deck would not run, as no navigation commands are given to get the different cards to display. I have included it here purely to give you an idea of what the structure looks like. If you are familiar with HTML at all, then the similarities will be instantly apparent:

```
<wml>
    <card id="First_Card" title="First Card">
```

```
    <p>
    Introductory Text.
    </p>
  </card>
  <card id="Second_Card" title="Second Card">
    <p>
    Get some data here.
    </p>
  </card>
  <card id="Third_Card" title="Third Card">
    <p>
    Display a result here.
    </p>
  </card>
</wml>
```

Limited Input Facilities

Wireless devices generally do not have the same input facilities as their wired equivalents; that is, they lack QWERTY keyboards and have mouseless interfaces.

WML addresses this issue, as well. The elements that are used in WML can easily be implemented so that they make very small requirements on the use of a keyboard. The use of decks and cards provides a navigation model that uses minimal navigation between pages, guiding the user through a series of cards instead of forcing them to scroll up and down on one large page.

Soft-buttons, or user-definable keys, are also supported by WML in order to provide the service developer with a means to couple desired actions to vendor-specific keys.

Limited Memory and CPU

Wireless devices are usually not equipped with large amounts of memory or computational power in comparison to desktop computers. The memory restriction applies to RAM as well as ROM. Even though it is likely that more memory and more powerful CPUs will be available in the near future, the relative difference will most probably remain.

WAP handles these restrictions by defining a lightweight protocol stack. The limited set of functionalities provided by WML and WMLScript makes it possible to implement browsers that make small claims on computational power and ROM resources. When it comes to RAM, the binary encoding of WML and WMLScript helps to keep the amount of RAM used as small as possible.

Limited Battery Power

The final limitation in mobile communications devices today is the operating time. The battery power defines the amount of time the device can be used. Even though battery technology is getting better and better, and the radio interfaces are tuned to consume less power, there are still limitations here.

Access to wireless services increases the utilization of the network interface, and so the power consumption also increases. The only way to solve the issue is to minimize the bandwidth needed by keeping the network utilization as low as possible.

THE BUSINESS CASE FOR WAP

When using the Internet from a desktop computer, it is very easy to find new and hopefully interesting services by using search engines, clicking on links and banners, typing in URLs recommended by a friend, and so on. We have the big screen, a full-sized keyboard and mouse, speakers, and a fast modem. We can simply sit down, surf the net, and enjoy the experience.

With WAP, it is very different. While we are on the move, we don't want to have to go and look for the services we want. We just want the information as quickly as possible, without all the bells and whistles.

This requires an approach to the mobile Internet that is quite different from the one we have become used to, with the graphics, sound, Flash movies, and so on. Instead of using advanced search engines and full-fledged portal sites, mobile users want small portals providing access to the services and information that they need, whether those portals are for business or pleasure.

This opens the way to new opportunities for companies that either understand the customers' needs very well, or that can personalize such portal sites to meet the demands of each and every customer.

WAP Services

So, what kind of new opportunities are there? What kind of services do the existing users want? While WAP is right now still looking for the next killer app, most of the online services that we are used to today can be of interest in the wireless community as well. The key issue in successfully launching these services is *usefulness*. If the usefulness factor is not high enough, then the majority of users will just ignore the service.

Bearing in mind the entertainment factor, the usefulness of a game might be very high in a certain percentage of the population, as demonstrated by the remarkable success of a company in Japan that provides cartoons for WAP phone users to send to each other at a premium rate.

We also need to remember that the vast majority of the public is not very familiar even with basic Internet services today. However, some examples of useful mobile services are in the following fields:

- ▼ Banking
    - Accessing account statements
    - Paying bills
    - Transferring money between accounts

- Finance
    - Retrieving stock and share prices
    - Buying and selling stocks and shares
    - Looking up interest rates
    - Looking up currency exchange rates
- Shopping
    - Buying everyday commodities
    - Browsing and buying books
    - Buying CDs
- Ticketing
    - Booking or buying airline tickets
    - Buying concert tickets
    - Booking theatre tickets
- Entertainment
    - Retrieving restaurant details
    - Looking up clubs
    - Finding out what is playing in what cinemas
    - Playing solitaire games
    - Playing interactive games
- Weather
    - Retrieving local weather forecasts
    - Looking up weather at other locations
- Advanced phonebook management
    - Updating a personal phonebook
    - Downloading a corporate phonebook

WAP also opens new possibilities for service and content providers, since they do not necessarily have to come to an agreement with a specific operator about providing services to their customers. This offers several benefits:

- You only need to create a service once, and it is then accessible on a broad range of wireless networks.
- You can build and address new market segments by creating innovative mobile value-added services.

- You can keep existing customers by adapting current Internet services to WAP.
- Creating a WAP service is no harder than creating an Internet service today, since WML and WMLScript are based on well-known Internet technology.
- You can continue to use standard tools like ASP or CGI to generate content dynamically.
- You can continue to utilize existing investments in databases and hardware that are the basis of existing Internet services.

The following are some example WAP applications:

- **123Jump (http://www.123jump.com)** A selection of stock data and news, all via WAP
- **1477.com (http://1477.com)** WAP/Web development services
- **2PL World-Wide Hotel Guide (http://wap.2pl.com)** A worldwide hotel guide, accessible in multiple languages via a WAP-enabled device
- **AEGEE-Eindhoven (http://wappy.to/aegee)** A Europe-wide students' association whose goal is to allow all students to integrate and learn about each others' cultures
- **Ajaxo (http://www.ajaxo.com)** A WAP service for Wireless Stock Trading from any WAP-enabled device.
- **Aktiesidan (http://mmm.aktiesidan.com/servlets/aktiesidan/)** A Swedish stock-market monitoring service, all WAP-enabled
- **Amazon.com Bookshop (http://www.amazon.com/phone/)** Amazon.com has launched this WAP portal (HDML-based) for browsing books
- **Traffic Maps (http://www.webraska.com/)** A French service that monitors and shows the latest in traffic news via maps

You can have anything you like on a WAP site. People have already set up WAP sites that deliver all sorts of personal content, from daily shopping lists to contact lists.

For example, you can put your entire contact list on a WAP page and use any WAP-enabled phone to access the names and phone numbers. (You can write a very simple password screen if you are concerned about security—we will look at this later on.) You could also update and add to your contact list from any Web browser on any PC, which means you can enter the information using an ordinary keyboard. As any mobile phone user can tell you, this alone will make the mobile phone itself far more useful just as a phone!

The following sections look in a little more detail at a few types of useful WAP services—services that are based on the user's location, those that offer customer service, and those that can be used within a business.

Location-Based Services

Obviously, services that can be used while users are actually mobile are best suited for the mobile Internet. Location-based services are services that know exactly where you are located in the world and can provide you with information that is relevant to your position. Traveling in a strange city? Feeling hungry for Chinese food? From your normal WAP menu, click on Restaurants. The gateway interrogates the phone network and determines which radio cell you are connected to. It then provides you with a list of local Chinese restaurants for menus and prices, and even gives you a map of how to get to the one you choose.

Or how about a service that displays the current physical location of different types of public transport. Let's say you get to the bus stop and you are late for a meeting. You need to find out if the bus has just left the stop, or is ten minutes late. At the bus stop there's usually a timetable, but this bus stop also has a unique number printed on it. You access the public transportation site from your mobile device, and type in the unique number. The Web server at the other end then knows exactly where you are and can display the exact time of arrival of the nearest bus heading in your direction, because the bus has a GPS (Global Positioning System) on board. An application like this could be enabled today in virtually any modern city in the world without having to wait for any new technology.

Or how often have you been stuck in a traffic jam? Do you stay on your current route and hope that the traffic clears up in a few minutes, or do you try to take another route? In England, most of the major motorways and traffic routes already have traffic cameras that are used by traffic control and emergency services personnel. By telling such a system roughly where you are (for instance, by telling it the motorway number and the nearest junction), the system could bring up a graphic of the traffic ahead and directions for the most rapid route to where you want to go if there is a traffic jam.

Such a system already exists, in fact, and stylized maps of the cities' main routes and traffic black spots can already be obtained for a number of cities in Europe. The corporate site can be viewed at http://www.webraska.com, and it is a good example of an application that has already been created for WAP and that demonstrates its usefulness.

Customer Care

Customer care is another place where WAP services can be of use. Substantial amounts of money are spent on voice call centers, through which people ask questions about their bills, or the features of a service are explained.

Many companies have successfully launched Web-based customer care services, as well, allowing users to access support data online. These online services can be designed to speed up the process at traditional call centers by, for example, having the user fill out a questionnaire to pin down the problem before the customer-care operator is contacted.

This online approach, however, does not solve the problem entirely, since customers generally do not have access to the Web when they are on the move. With a WAP-based customer care service, the customers would be able to select from multiple choice menus to pin down their problem and get help whenever they want, without having to spend a substantial amount of time waiting for their call to be answered.

In addition, the WAP Forum is looking at several technologies that will enhance mobile value-added services, such as persistent storage, the use of smart cards, external interfaces, billing, data synchronization, and user-agent profiles. Two areas that will have direct impact on the services we will see in the future are *push technology* and *Telematics*.

Push is the ability to send text messages to a given phone. Just as you can request a daily news digest to be delivered to your PC via the Web as an email, so too can you request messages to be sent to you on various topics, from daily horoscopes to stock results.

Telematics is the technology of automotive communications that combines wireless voice and data to provide location-specific security, information, productivity, and in-vehicle entertainment services to drivers and their passengers. These can be such services as dispatching an ambulance to a driver in an emergency situation, or sending a roadside assistance service to a driver whose car has broken down. Other services include the delivery of navigation assistance and real-time traffic information. This is letting the user know about his environment.

With a location-based system, we can also let the environment know about the user. For example, McDonald's will know when a telematics user matching the profile of a McDonald's customer is within the proximity of one of their restaurants between 11 a.m. and 2 p.m. McDonald's will then be able to send a message to that person. Of course, the user may have the ability to decline to receive such messages, but they will most likely be rewarded (perhaps with discounts) if they agree to receive them.

The word Telematics itself was first used by Mercedes-Benz to describe their automotive communications technology, and has since caught on.

Wireless Employees

Now let's imagine that you work at a road construction company, building and repairing hundreds or perhaps thousands of roads. Typical projects are discussed in the hundreds or even thousands of employee-years. Your organization learned some time ago to make use of advances in computing technology by delivering real-time access to information via mainframe terminals and later Windows applications on employee desks or on workshop floors. This opened up existing databases to improved reporting, charting, and other user interface features.

Managers and site foremen could access parts inventories, repair schedules, shop budgets, and other useful information in order to plan work-crew schedules and employee tasking.

It was just another small step from there to Web-enable various mainframe applications. This information on the Web can be shared with parts suppliers and contractors, greatly reducing ordering times and costs.

However, out of perhaps 10,000 employees and contractors, only about 500 are actually interacting with the databases. The remainder of the employees continually fill out paperwork, issue reports to their managers, or manually key in data when they return from working in the field.

Imagine if all of the other 9,500 employees actively involved in laying tarmac, installing electrical cable and conduits, and building drainage systems could access or edit data when they actually need to. A small, inexpensive device could be given to each employee based on his or her requirements.

Some employees require handheld devices with built-in barcode scanners, others require keypads, and others require simple digital displays. WAP allows a suite of client applications to be built, reusing existing server applications and databases. In addition, these applications can be dynamically downloaded and run on any of these devices. If a cable installer realizes that 500 more feet of a specific type of cable are required, she selects the Order Cable menu option from her WAP-enabled phone. If someone installing a storm drain wants to know which pipes or cables are supposed to run through an associated conduit, he enters the query into his PDA and retrieves either data or image information.

In any industry that involves employees stepping out of their offices to complete a job, wireless applications can markedly increase productivity and your ability to remain competitive.

Why WAP?

Some critics have pondered the need for a technology such as WAP in the marketplace. With the now entrenched and widespread use of HTML, is yet another markup language really required? In a word, yes. WAP's use of the deck-of-cards model and its use of binary-file distribution ties in and works with the display size and bandwidth restrictions of typical wireless devices in a way that HTML never could.

In addition, scripting with WMLScript gives support for client-side user validation and interaction with the portable device, which helps to eliminate round trips to remote servers.

WAP's biggest business advantage is the prominent communications vendors who have lined up to support it. The ability to build a single application that can be used across a wide range of clients and telecom carriers makes WAP virtually the only option for mobile handset developers at the current time. Whether this advantage will carry on into the future depends on how well the vendors continue to cooperate (via the WAP Forum) and also on how well standards that are agreed upon and issued by the WAP Forum are followed.

The reason for the business interest in WAP isn't hard to figure out. According to market research company IDC, the revenue from the European mobile commerce (m-commerce) market alone will rise to $10.8 billion, or 13 percent of the mobile phone companies' revenue, by the end of 2003. According to Forrester Research, 90 percent of large media companies, retailers, and banks in Europe are developing online services for mobile phones in preparation for the expected demand.

Content providers have been won over to the technology by the belief that with nearly 130 million European subscribers, mobile telephones that have been WAP enabled will rapidly gain mass-market acceptance.

Unfortunately, although it is still very early in the game, vendor toolkits are already offering proprietary tags that will only work with that vendor's micro-browser. Given the history of the computing industry and competition in general, this was only to be expected. However, further differentiation between vendor products and implementations may lead to a fragmented wireless Web.

WAP is a young technology that is certain to mature as the wireless data industry matures, but even as it exists today it can be an extremely powerful tool in every software developer's toolbox.

THE FUTURE OF WAP

The future of WAP depends largely on whether consumers decide to use WAP devices to access the Web, and also on whether a new technology comes along that would require a different infrastructure than WAP.

On the consumer side, the factors largely involve the limitations of WAP and of handheld devices: the low bandwidth, the limited input ability, and the small screens all require users to adapt from their regular Web-browsing expectations. The bottom line is that WAP is not and can never be the Web on your mobile phone—WAP is great as long as developers understand that it's what's inside the applications that matters, and the perceived value of the content to the user. The browser interface itself, while important, will always be secondary to the content.

On the technological side, it is true that mobile phone data speeds will get faster, and this may require a different infrastructure, which will require different handsets to carry the technology. The new challenger is General Packet Radio Service (GPRS), but the rollout of newer technologies will take at least a couple of years. And although gadget freaks, like me, will probably get one right away, there will be a lot of people who will remain quite happy with their legacy phones, and who will resist spending money on something that they can't see as being immediately necessary.

On that basis, you can estimate that WAP, even as it is today, will be around for a minimum of three to four years. Given how quickly things are changing because of the introduction of the Internet, this is an incredibly long time.

Also, changes or enhancements to physical technologies will not necessarily require completely different protocols. In terms of WAP, the transport layer or protocol may have to change, but the content that will be transmitted will still have its own unchanged format—WML. That means WML developers will, in real terms, be virtually unaffected by any hardware changes, and can continue to develop applications regardless of any hardware changes.

Web browsers for the PC will also soon come with the ability to view WAP pages. There is already at least one product available for free, called the WAPaliser, which allows you to view WAP pages from your Web browser. What this means to the average user is that the instant information access available to mobile users can be combined with the rich content of the Internet. You will simply have two windows open on the PC—one for the traditional content and the other for the WAP content—thereby having the best of both worlds.

Some major software-development products are already including the ability to create WML decks and cards in a WYSIWYG fashion. Macromedia's Dreamweaver already has such a plug-in, and it won't be long before the other players follow suit.

With this much investment of time, effort, and expenditure by a lot of major names in all areas of software and hardware, it is highly improbable that WAP will become obsolete anytime in the near future. The mobile Internet with WAP will probably change from the way it looks at this time of writing, but what has already been developed is a very good start.

CHAPTER 2

What Makes a Good WAP Application?

Let's take a look at what is really needed and wanted by users of WAP devices. They want to be able to solve a real or perceived problem. They want to be able to do so quickly and easily. If they can also have some fun at the same time, they will come back and use that solution again and again, and they will recommend it to their friends and colleagues.

There is a very simple equation that can be applied: The perceived value of the application has to be greater than the perceived cost. If this holds true, the application or service will be used. If it doesn't, then the service may be used once or twice and then dropped. Just as importantly, the user will give negative feedback on the service to friends and colleagues.

This means that if it takes the user five minutes of struggle with an unfriendly application to find a traffic report, which is then found to be inaccurate, that user will never use the application again. He or she will thereafter simply switch on the radio and get the data from there. If, on the other hand, the user can get up-to-date and relevant information in 30 seconds, they will continue to use the application in the future.

You must remember that a user doesn't care about your company's business model, policies, investments in new technologies, or anything else. If the application is easy and relatively cheap, the user will come back. If the application is not easy or relatively cheap, then the user won't come back. It's that simple.

The purpose of this chapter is to look at various ways in which we can help make the user's experience a fruitful and pleasant one, so that the application will continue to be used.

THE USER'S POINT OF VIEW

Before you can design a good WAP application, you need to understand who will be using the application and why. The person using a WAP device is in quite a different state of mind from, say, a person sitting down to use a PC or making an ordinary phone call.

Most obviously, they are mobile, which means that they may or may not be sitting down; they may have all sorts of external distractions; they may be driving in traffic or standing in a store doorway.

They are often traveling with a specific goal in mind, such as going to a meeting, or returning from a meeting that didn't go well.

Most importantly, though, they will be thinking, "What access can I get to the information I want?"

You are going to present them with questions and information on a very small screen that they will have to follow correctly in order to get the information that they need.

It is also good to keep in mind that we are impatient, we ordinary people. We want to get to the content we want as quickly as possible. We don't want to have to wait. We don't want to register for a site before we know what value we will get from it. We don't want to have to figure out what to do next—it should be clear.

Ease of Use

The "average person" does *not* read the manual, particularly for something like a telephone. We try to figure it out for ourselves. If we can't, we get frustrated and tend to leave it altogether. That means that if your application is not *truly* intuitive, then people will get stuck, and you will lose them forever. The user will not be interested in solving the problem of how to get the application to work.

It is also important to note that at this early stage of WAP development, people are buying mobile phones so that they can use them as phones. The fact that they have this "WAP thingie" installed is usually not a primary consideration in the buying process. Most marketing departments over-hype their products to the point where buyers have become very cynical. Most people find it hard to believe the sales pitches about "surfing the Internet" from a WAP phone, and very correctly so.

Another ease-of-use factor is the difficulty of getting to a URL. The average person will only ever type a URL into a WAP phone once. Only once. Because typing a URL into a WAP device is so awkward. If what users want isn't already on a menu, they probably won't bother.

For a developer, this means it is vital that you get your site listed on a main WAP portal site so that it is easily accessible to the user from one of the basic menus. The closer to the top of the menu hierarchy you can get your site listed, the better. You can assume that for every additional step down the menu hierarchy, you will lose at least 50 percent of the users that have made it that far. If 10,000 users hit the first menu, only 5,000, *at most*, will make it to the second menu. Only 2,500, *at most*, will make it to a third menu.

In summary, if you have an application that is easy to use and provides a genuine solution to a perceived problem, and the users can find it, then people will use it often and recommend it. If you have an application that provides a genuine solution to a perceived problem but it is *not* easy to use, people will stop using it when the perceived cost of the application is greater than the perceived value.

Designing for Users

The first thing to decide is why the user is going to be using your application. To get the local weather report? To get a current stock quote? To remotely read office e-mail? To get the address and phone number of a local restaurant in order to make a booking for that evening?

This is usually pretty obvious, as it is generally the purpose of the application, and the reason it was written in the first place. However, if the user's goal and the application's goal are not the same, there may be a problem. Suppose the user wants to know the weather for the coming weekend in order to plan for a family picnic, and your application only provides the weather for the next 24 hours. In this case, your application is not providing the solution to the problem and thus will be remembered as "not useful." What most often happens is that the user will find an alternative "better" service and use that one exclusively.

The next thing to decide is exactly when the user is most likely to be using your application or service. First thing in the morning over breakfast? In the car during rush-hour traffic? During the lunch break over a cup of coffee? This is extremely important to know. If the application is likely to be used when the user is relaxed and has time (say for an application that delivers personal horoscopes), you can offer more data, or provide a richer interface, or links to related services or applications. If the user is in the car and wants to know the condition of the traffic so he can pick a route to work, he will not want to be scrolling down long lists of choices or want to read additional data. He will just want to know, "Should I take Route A or Route B? Route B. Good. (Click)." If the user is likely to be using the application at night, is the screen clear and the font size set to large so as to be easy to read in bad light?

Critical, Useful, or Fun

You need to establish whether the information that you provide is critical, useful, or just fun or "nice to have."

An example of an application that provides *critical* information is an automatic notification message that alerts the user to the fact that the price of specific stocks or shares has dropped or risen to a predetermined level or by a predetermined percentage so that they can take immediate action.

Useful information might be access to the local phone directory, or to the news, weather, or traffic reports.

Fun or "nice to have" applications might involve access to online gaming sites, reminders of family birthdays, or details of the latest items on sale in the local supermarket.

Depending on the user and the situation, a useful piece of information might become critical. For example, knowing that a partner's birthday or your wedding anniversary is a month away is useful. Knowing that it is today is *critical*.

Similarly, there will be times when a critical piece of information becomes just "nice to have," like a stock quote on a weekend.

"Solutions" Versus "Features"

You probably know all kinds of applications that contain lots of features, but at the cost of usability. With very few exceptions, most off-the shelf software packages sold these days could have actually benefited from being held at an earlier level of release.

With PC software, the drive has always been to make it bigger and add more features, more functions, and more options. There are good business reasons for this. If a company isn't seen as having the latest, the biggest, the best, then they risk losing sales to the competition. Also, bringing out a new release once a year gives customers, both existing and new, the comfort factor that the software company still cares and is supposedly listening to the requirements of the customers.

However, it has been estimated that no user of Microsoft Word, for example, ever uses more than ten percent of the total functionality of the program, and that's after extensive training.

The benefit of larger feature-rich programs for PCs can be debated, but with WAP, bigger is not better. In fact, it is a positive liability. Users are not only mobile, but they are often paying by the minute for connection time. They aren't wanting much except to get online, get the data, and get offline again.

WHAT ARE THE WAP MICRO-BROWSER ISSUES TODAY?

WAP itself is a standard, which mercifully is being adhered to by the different manufacturers. That's the good news. The bad news is that the presentation layer, which is the one that determines exactly how the WAP pages are going to be displayed to the user, comes in several different flavors, depending on the phone manufacturer and the micro-browser package.

The precursor to WML, HDML, is also still very much in use in the United States. HDML is Phone.com's proprietary language, which can only be viewed on mobile phones that use Phone.com micro-browsers. HDML is being replaced by WML all over the world—WML is already the standard in Europe and Japan, and most phones in the United States and Canada will also accept it. However, Phone.com promises to continue to support HDML.

Adding to the confusion is the fact that new phones are now starting to come out in larger and larger numbers from more manufacturers. These can each have their own "features" (for *features* read *differences*) in terms of display sizes and keypad layout.

WAP micro-browsers are constantly evolving; for example there are over twenty different versions of the WAP micro-browser for the Nokia 7110, one of the first WAP-enabled phones on the market. This means that different phones of the Nokia 7110 model could have one of twenty different micro-browsers, depending on which micro-browser was current when the actual phone came off the production line. Unfortunately, this also means that applications can behave differently even on the same model of phone.

In addition, each of the major micro-browsers has their own code extensions, over and above the standard WML versions 1.1 and 1.2, which were agreed upon by the WAP Forum.

If you use the Phone.com micro-browser, you can do things like sending notifications, keeping bookmarks, and using cookies at the actual WAP Gateway server (a specialized piece of software that acts as the gateway between the mobile network and the Internet). With the Nokia micro-browser, you currently can't use cookies, which means that you have to find alternative ways of doing these things if you want to use cookies and you are targeting the Nokia micro-browser (in Europe and Australia).

However, targeting the Nokia micro-browser isn't much use if you are trying to reach American users, because the carriers in the United States only install and support the Phone.com micro-browser at this time.

The WAP market is currently in the same situation as the World Wide Web standards when they first arose. Going from text-only e-mail to graphics pages was an exciting time for developers, but only for the developers who backed the winning applications. Hopefully, more standards will emerge that will unify the offerings available to the general public; otherwise the present confusion will delay the emergence of true m-commerce.

The bottom line for developers is that you cannot guarantee the way the page will appear on any model of any phone. It fact, even the simulators used by developers for a specific model phone (more later) do not always have the same behavior as the actual phone itself once released to the market.

Now that I have pointed out all the terrible things about developing in this arena, let's see what we can do to reduce some of the confusion and start making usable applications that are transparent to users, no matter what micro-browser or phone they may be using.

Micro-browsers and gateways are currently used as follows:

Location	Browser	Gateway
North America	HDML	Phone.com
Japan	I-Mode, HDML	I-Mode, Phone.com
Europe and Australia	WML	Phone.com, Nokia

We will look next at the use of "generic" WML, but it is important to keep in mind that you should design the application so that it is easy and intuitive to use, regardless of the presentation layer.

Writing a Generic WML Interface

I have already mentioned that different micro-browsers can have different extensions depending on the developer's kit installed and used. So where is the common ground?

Standard WML 1.1 has quite a number of features, and all of the basic features will allow you to create the "decks and cards" that are the building blocks of any WML application. Any developer familiar with HTML can learn the basic syntax of WML in a matter of hours. So if we all just write our applications in the standard "generic" code, there won't be a problem, right?

The advantages of writing generic code are fairly obvious. You only have to write the code once, and the pages (or "decks") can be hosted in one location. The sites theoretically take less work to develop, they are faster to get to the marketplace, and they are easier to maintain.

In reality, though, it is not so easy to just write generic code. For one thing, just because a WML tag or syntax is part of the WAP standard does not mean that all of the phones and micro-browsers will recognize it, or work with it even if they do recognize it. A classic example of this is the <TABLE> tag in the Nokia micro-browser. I once spent over an hour trying to get a table with two columns to display in the Nokia 7110 WAP phone. No matter what I did or how I coded it, it would not display as a table. There were no error messages—the simulator would simply compile it and display each of the elements on a separate line. In desperation, I at last turned to the Nokia WML guide and eventually found a little line that said "Although you can use the <TABLE> tag, you can at this time only have one column in the table." Or words to that effect.

Which effectively means that you can't display a table on the Nokia 7110, no matter how nice your code is, or how standard your code is. The work-around is, of course, to format each line on the display separately, adding spaces and so forth, if you want to simulate a table on the Nokia 7110. But this doesn't help if you want to write a "generic" application. Generic standards say that a table is okay.

So while writing generic code sounds good in theory, the reality is that you can spend more time and effort writing a generic application that does not, deep down, satisfy anyone, than if you were to acknowledge the problem in the first place and design your application accordingly.

You also have to remember that there are many more, different phones being made and released all the time. Some of these are going to offer the same functionality, even though the styling is different. Some of them are not. That means the amount of code that is truly generic is currently shrinking, not growing as we would like it to be.

Because of the differences in the presentation layer, your applications are going to look different anyway, so you might as well take advantage of this fact and build applications that anybody can use, no matter what their micro-browser or gateway provider, and that are optimized (as far as possible) for each one.

The only alternative to writing generic code ("one size fits all") is to write for specific models of WAP device, and have the code available to the model of WAP device and version of micro-browser on demand. You could take the approach that you could write a different application from the ground up for each model of WAP device, and put each application on a different site or sub-directory.

Or instead of putting different versions of your application on separate sites, you could have one set of code with many IF-THEN statements. If the phone is a Nokia, do this; if the micro-browser is a Phone.com micro-browser, do that; and so on.

This approach can be very messy if it is not organized properly. You also have to draw the line somewhere; otherwise you would be doing nothing except adding to your code for each new phone and micro-browser version that comes out. However, it does offer the advantage that all of the controlling code is in one place, and individual options can be written as server-side include files. That way, different developers can be working on different presentation layers for the different phones without treading on each other's toes.

Another major advantage of this approach is that when a new user agent comes along, like a WAP-enabled Electrolux refrigerator, or a WAP-enabled PDA, you can add a whole new class of devices to your application with an additional IF-END IF block in your code. And believe me, new user agents will be coming along.

Targeting Your Market Micro-Browser

The days of the "one device" that will sit on everybody's wrist and handle all of our communication and information needs are still a little way off yet. Until then, each manufacturer is going to have a finite slice of the market. If your application can reach any phone and work in a way that is completely familiar to the user of that device, then it has a chance of being used by far more people than if it doesn't.

You can, however, decide to just target all Siemens users, or Ericsson users, or Nokia browsers, or, or, or. You could do this. After all, Siemens, Ericsson, and Nokia all do this already, as a matter of course.

There are definitely advantages of targeting to one device. For example, you can gain a larger share of a specific market and form strategic alliances with the manufacturers and telecom carriers concerned. Also, the users may be forced to use a limited number of gateways and, in some cases, may not be able to reach your application at all! (In France, for example, the major carrier is planning on limiting user access to a specific approved set of services or URLs unless the user pays for the "premium" service.)

There are currently differences between carriers in countries around the world. Notably, in Japan, the main carrier DoCoMo dictates to the manufacturers the exact specifications of the phones. The display has to be a certain size, in terms of rows and characters; there have to be particular buttons labeled in a certain way; and the micro-browser has to be able to do x, y, and z. In North America and Europe, the manufacturers tell the carriers what phones they are going to be producing.

For carriers, this is a small problem, as they already own the infrastructure. For Japanese developers, the standardization is great. For the rest of us who don't have such agreed-upon (or decreed) standards, it's not so easy for now.

HOW TO DESIGN A GOOD WAP APPLICATION

What makes a "good" application? One factor is the business model—the problem that the application is trying to solve. That can be anything from providing up-to-date sports news, to offering a full online purchasing system via a company extranet. The business model is limited only by your ability to perceive a user problem or requirement that a wireless application could resolve.

Any developer or software house will know that it is not possible to say what a "good" application *would* be. It is only possible to say what a good application is after delivery, by virtue of the fact that users are paying money for it. The massive sales of things like Teenage Mutant Ninja Turtles and Pokemon tells us that it is not always possible to predict the next "killer application." What we think users want and what they end up buying are often two different things. In fact, what users *say* they want and what they actually buy are also often two different things.

A recent study in Japan shows that while only 11 percent of prospective phone buyers say they will use the phone for entertainment, 25 percent of phone users actually do use the phone for entertainment. Interestingly enough, the figures are reversed for financial applications. Far fewer mobile users actually use their phones for financial transactions than say they will.

While I can't provide you with the secret of creating the next killer app to sweep the world, I can at least let you know the best way to design any application so that it will be usable and useful to the end user.

WAP is a different medium from the Web, in a class all of its own, and we can lay down some ground rules for designing an application that fits the medium.

When building a web site in HTML, there are all sorts of rules and guidelines available: keep graphic images as small as possible, pastel colors and crisp high-contrast text works for mass market consumers, white text on a black background and loud colors do not inspire confidence, discrete animation works, lots of "in-your-face" animation tends to be off-putting, and so on.

With WAP, there are different rules. Firstly we have content limitations:

- ▼ **Text**   Introductory paragraph, or page? You have about six lines of fifteen characters per line. You will find unparalleled joys in being able to find a single six-letter word that you can use as a link to "Economic Development Agencies." The basic rule is, "Keep it short."
- ■ **Images**   Right now, you can include whatever you like, as long as it's black and white (or gray on some phones) and will fit into a square about 1.5 inches on a side.
- ■ **Colors**   Forget it. Phones with color screens will be arriving, but you then have to worry about bandwidth (because color images are larger), and you would also need to make an image and content that is also easily visible in black and white, so that legacy phone users can see it.
- ■ **Frames**   Frames don't exist as a part of WML at this time, and as there is virtually no screen size to subdivide into frames, this is no real loss.
- ▲ **Flash content**   You have the same problems with Flash content as with images, plus I don't think a Flash plug-in exists yet for WAP browsers. Flash content also tends to be fairly high in bandwidth consumption.

Then we have device limitations:

- ▼ **Keyboard and mouse**   These are only available if you buy special add-ons for specific phones, and then it isn't really a "mobile" anymore, is it? In any case, you can't assume many users will have them from a developer's viewpoint.
- ■ **Speakers**   This means audio content, which means bandwidth. A PC user can get streaming audio while surfing. I quite often listen to U.S. radio stations while I am surfing or researching on the Internet. And because I can easily multitask and have multiple browser windows open, I get easily detectable audio cues when a file transfer has completed or an error occurs. Until speaker technology develops, I cannot both look at a mobile devices screen and listen to the earpiece at the same time. And with a 9.6k data connection, I don't think I want to pay for the audio file download time either.
- ▲ **Desk and chair**   While it's true that many users will make use of their WAP phone while sitting down somewhere when they have a few spare moments, the whole idea of a mobile phone is that it can be used while actually mobile.

The Application-Design Process

So how do we design an application that is going to work well within the restrictions of WAP devices? Let's take a look at the process that we need to go through.

1. Who is your customer?
2. What micro-browser will you design for?
3. What phone will you design for?
4. What problems will the user run into?
5. Solve each of the problems you have visualized.
6. Draw the screens for the application.
7. Test the mock-up.
8. Write the pseudo code.
9. Write the prototype.
10. Test the prototype for usability.
11. Fill in the gaps.
12. Test the final product.

(Incidentally, while some of these steps apply to the good design of any application, this discussion applies specifically to mobile devices.)

1. Who Is Your Customer?

First of all, you need to have a very clear idea of exactly who the end users will be for the application that you want to build. Are they going to be young or old? Male or female? Wealthy or not? Blue collar or white collar?

2. Which Micro-Browser Will You Design For?

Having established your profile of prospective users, the next major question is which micro-browser you are going to consider the most important, and in which sequence you will develop them.

This will depend largely on where your users live. If your end users are primarily in the United States, you will be developing primarily for the Phone.com micro-browser, as it currently has 100 percent of the market there. However, if you are targeting Europe, the choice isn't quite as clear-cut. Although Phone.com claims to have a larger market share in Europe than any of the other micro-browsers, Nokia has large brand recognition.

This may well require a little more research on your part as a developer before you begin the actual coding process.

3. What Phone Will You Design For?

What phone. The nightmare begins. On a small-screen Nokia you have a roller that scrolls the screen up and down, and pressing the roller acts as a button click. On a Siemens phone you have little rocker buttons instead of a scrolling wheel, the screen dimensions are totally different, and the Select key is a button on the other side of the phone. And so on.

It is only by talking to the retail outlets that you will really find out what phones are already out there and in what numbers. You have no choice but to go with the consumers, choice. The Nokia 7110 has probably the smallest displayable area in terms of characters of any WAP-enabled phone on the market, but the styling of the phone itself is irresistible.

If you are designing a general, horizontal-market application, it is best to assume that the screen size will be the lowest possible. This will not affect its appearance on a larger display, and may even enhance it. Conversely, you won't score any points with users if a large proportion of them using smaller displays see a lot of unnecessary word wrapping.

It is quite a different scenario when you are dealing with a vertical market, and you can design for a specific phone or range of phones quite happily. In this case, you can even form a strategic alliance with one of the phone carriers or manufacturers so that by promoting the application you also promote the specific phone to use it on.

It is only a matter of time before somebody starts promoting the "stock-market phone" or the "supermarket phone," where the application has been optimized for a given phone to the point of complete integration. This kind of integration makes sense if you have a very clearly defined market.

This kind of integration also makes it very easy for the user to use the application. Press one button to enter the supermarket (dial, get online, and go to a specific URL), press another button to see special offers (updated dynamically—check back every hour from your desk or easy chair), press this button to order and that button to pay (automatically adding the cost of the items to your telephone bill and getting the goods delivered right to your door). If you kept the basic items on the main menu (for example, bread, milk, eggs, sugar, etc.), and personalized the menu for that customer based on the items they have ordered in the past, you might have a winner of an application.

4. What Problems Will the User Run Into?

One major point of failure for any software application is failing to understand the problems that ordinary users will run into in the normal use of that application. Choices that are confusing, buttons that either aren't obvious or clearly labeled—these are minor things in themselves, but they will force users to stop and think about how to use the application rather than letting them think about what they want to accomplish. This eventually causes people to stop using the application.

One way to handle this is to imagine each of the different times that a user will be using the application, and what the scene might be. For example, people are likely to want to use their WAP-enabled phones to look up the local taxi telephone number between 2:00 and 3:00 A.M. What problems are users likely to have?

The most obvious difficulty is that it will be dark at this time, and the user may not have a lot of light to see the phone's screen. That means the lettering should be bold and as clear as possible. Also, the user may well have been drinking, so any choices they have to make should be very uncomplicated.

The user could also be with one or more other people who may well be talking, joking, and otherwise distracting the user from the application. That means users may need to be shown exactly where they are in the application at all times, so they don't lose their place.

Each of these user problems is a specific problem that you, as the developer, may never have thought of. You do need to specifically visualize the circumstances occurring when the user actually uses the application.

5. Solve Each of the Problems You Have Visualized

Solving potential user problems may be as simple as rewording a question or instruction, or it may require more innovative changes.

In the case of the taxi service, you could decide to have the application check the time of day. If the time is between 10:00 P.M. and 6:00 A.M., you could make the wording bold and double-spaced for legibility; otherwise you could have it display "normally" during the day so as to get more data on the screen.

If you run into a problem that you cannot think of a resolution for, at least you are aware that the problem exists, and you can keep an eye out for ways that you could resolve it as technology or coding standards change.

6. Draw the Screens for the Application

Armed with all of the information from the previous steps and the specifications of the application itself, you can now work out what screens are necessary and in which sequence they should appear. Your best tools at this point are a pencil and paper. If you can't get it so that it looks okay on paper, then it will never work for the user. Each screen should resolve one specific question that is necessary for the application to produce a result. Remember, there is no room for "fluff" in a WAP app, unless it is supposed to be an entertainment application.

The clearer you can make the screens and the flow of user direction from one screen to the next, the better you will be able to code your application, and the more the user will use it. Remember to include all of the solutions for the problems that you came up with in step 5, so that they don't get lost and have to be re-solved.

7. Test the Mock-Up

Once you are happy with the mock-up you have created, show it to a few people at different levels of technical proficiency to get their feedback. Were they confused on one or more screens? Did they feel that they had to take more steps or press more buttons than were necessary? One good feedback comment at this point can save you a lot of wasted time and effort later on.

8. Write the PseudoCode

Now you are ready to write the pseudocode that takes into account all of the points above. I have found that writing pseudocode is the most overlooked part of the creation of any application. It is a very simple step, but it accomplishes two major goals. Firstly, it allows you to spot problems very early on, and work out solutions "in the rough" as you go. Secondly, you can get the whole of the application or the module "out of your head" and onto paper, so that you can see the broad picture. This is the time to include dummy IF statements that include all of the micro-browser and phone points you have decided on.

There have been several times when I have been writing pseudocode in plain English, following a particular line of logic, and realized "hey, this won't work" because I had overlooked a limitation of the language I was coding in. In pseudocode the resolution is easy—just cross it out and start again. There is nothing worse than coding yourself into a corner, and ending up writing some extremely convoluted code to get out of it. This is all too easy to do if you just start writing code instead of doing the pseudocode first.

An additional benefit of writing pseudocode is that you can talk somebody else through it easily because it's usually just plain English. That way you can see whether your assumptions are correct and whether the solution you have come up with actually holds water. You can also keep the original notes to refer back to six months later when a requirement changes. (Which it will. They *always* do, trust me.)

9. Write the Prototype

The next step is to put together the application "skeleton" so that there are real links between cards and decks. This is real code, but with most, if not all, of the application functionality left out. The navigation structure should all be there, so that menu options and choices are all there and working. The actual cards and decks are all created, even if they have no content at this point.

This will allow you to test the usability of the application (see how the screens will tie together on the actual phone, determine whether the mock-up is really usable and intuitive, etc.). You can insert dummy code and "stub" functions wherever you like, to simulate a fully working application.

> **NOTE:** A stub function is a function that simply consists of a "return" statement for the time being. You can put a function call to a stub function in your program where you know a call will be needed—this allows you to complete the main program without having to work on the peripheral functions first. When the time comes to get the function to work, you can add the code in the stub function where appropriate, and the main code does not need to be touched.

In WML, you would insert cards and decks with dummy statements like "the date goes here" in them. You can also put in dummy cards and decks for each of the phones and micro-browsers you are targeting, with all of the IF-THEN-ELSE functionality in place.

Providing you have followed the preceding steps, writing the prototype application can be done very rapidly. If you find that you are having trouble writing the prototype, you should go back to the design stages to see where the problem lies, and resolve that before moving forward through the steps again.

10. Test the Prototype for Usability

For the prototype testing, you should try and get as many "typical users" as possible. Obviously the "typical user" is going to vary, depending on the target users that you picked in step 1.

Get your test users to pull the prototype to shreds for every combination of phone and micro-browser you have included. You should treat each combination as a separate "virtual application," with its own presentation rights and wrongs.

At every testing stage, the idea is to find out what is wrong with the application so that it can be fixed *before* the application is released to the public. With no real code in the application to go wrong and distract the testers at this point, you can concentrate fully on the important things for a mobile application: ease of use, ease of use, and, finally, ease of use.

11. Fill in the Gaps

Once user testing has validated the prototype, the rest is actually quite simple. You have your framework all written, and you have worked out how you are going achieve the application's functionality. The writing of the final application is then the purely technical activity of converting the pseudo code to actual code.

When you wrote the pseudo code, you were concentrating on the problems of functionality and ignoring the code syntax. When doing the final conversion to real code, and filling in the gaps, you can concentrate solely on the code syntax and getting the code to work. The major problems and function algorithms have already been solved in the pseudo code.

It is at this point that you write the actual code for the different presentation layers, and replace the dummy statements and layouts for each phone or micro-browser with the live code. Armed with copies of the relevant reference manuals and phone simulators, this becomes almost a mechanical task.

12. Test the Final Product

You have to test, test, and test some more. The use of simulators is vital, but unfortunately the simulators do not always reflect the actual behavior of the phone itself. More and more phones are arriving on the market all the time, micro-browsers are being revised, and the standards for WML are being upgraded, all of which makes a developer's life very hard indeed. Just do the most conscientious job that you can.

You are not going to get it right for all presentation layers, for all phones, 100 percent of the time, because you are always shooting at a moving target. Unless you are a major development company, it is unlikely that you will be able to go out and buy 150 different models of mobile phones just so that you can do testing, and even if you are, you have to draw the line somewhere.

This is not a reason for skipping testing, but don't get overwhelmed by the numbers of variables out there. If your application works on 95 percent of all phones currently available and new phones as they are released, it would be more than acceptable.

Keep testing and usability going as continuing actions. Your application can always be better than it is.

Common Design Mistakes

There are several common design mistakes that are often made when creating a WAP application, and these can easily be found in many existing applications. They boil down to application complexity, applying the wrong development model, and failure to do usability testing.

Application Complexity

Complexity is just the opposite of ease of use. If the application is complicated to use, or if users have to be rocket scientists to figure out how to get the information they need, then your application will lose. Users will not use the application, and the whole thing will dry up and blow away.

Wrong Development Model

Because of the underlying similarity of the technology to the Internet, and because the application is being developed on a desktop PC or Mac, it is all too easy to forget that a WAP-enabled device is a different platform and medium altogether. The same rules do not apply.

The only crossover is the information or data the user wants. If you try to make people use a mobile phone for browsing instead of for data access, the application will eventually fail.

No Usability Testing

Many of the simple usability problems can be easily spotted and removed at the earliest stages of the design and development process. If this is not done, then many very simple issues can be overlooked until it is too late. Discovering that users won't use your completed application is the wrong time to find out that there is a major problem in your user interface.

The bottom line is that an application with a potential market of 5,000,000 user minutes per month might only gain 50,000 user minutes. At 10 cents per minute, the difference of $495,000 per month makes it extremely viable to spend extra development time and effort on design and usability testing before a public release.

CHAPTER 3

The User Interface

We have looked at the application-design process as a list of general dos and don'ts, so now we will look at the specifics of a user interface for a WAP device. This consists of screen layout and menu design considerations primarily, but we will touch on other aspects as well.

USER INTERFACE BASICS

A *user interface* can be roughly defined as the set of routines and procedures that a user has to follow when working with a device, whatever it is, in order to use that device as intended.

User interfaces do not only apply to computer software. Anything that is used in some way has a way of being used. This is commonly called *design* or *style*. It's all part and parcel of the same thing, however. Even vegetable peelers differ between brands. The one that feels comfortable when you are actually using it is probably the one that has had the most investment in terms of design, or user interface.

Any user interface has to be *learned* by the user. Most people in the developed world can operate a television and any associated remote control fairly easily, and it is pretty much taken for granted that this is so. However, working a device like a radio or a TV still requires a basic set of experiential reference points. If the reception isn't good, we know that we need to adjust the aerial or complain to the cable company, depending on how you receive the transmissions. Even adjusting the aerial requires an understanding that the aerial needs to be positioned properly to catch these invisible signals.

Now, hands up if you can program your VCR easily. I still have trouble with mine, and I have had the same machine for over five years. In fact, the VCR is almost the archetypal example of a badly designed mechanical user interface.

So, if the user always has to learn an interface, we need to try and make the learning curve as easy as it can possibly be.

When people are using a mobile phone, they are working on the basis of their previous experience with telephones. It doesn't matter that the phone can access the Internet or act as a calculator, alarm clock, or whatever. If it looks like a mobile phone and acts like a mobile phone, then the laws of association say that I, and most users, will regard it as a mobile phone, and not as a mobile computer or anything else.

This is probably why there is some difficulty in working out an interface strategy for WAP devices in general. The brain says the device is a web micro-browser, but the eyes say it's a telephone (or a PDA, or a refrigerator, or whatever else it might be). This means that any user interface that will be used on a mobile phone must be designed for users who are primarily thinking of the device as a mobile phone.

To show how important some of the design considerations of a user interface are, here is an example from the car industry. Lotus Cars developed the "active suspension" system now commonplace on many modern cars. The original design goal was to be able to drive over a brick at 60 miles per hour and have the driver think that he had missed it completely. This goal was actually achieved—there was only one problem. Because the ride was so smooth and comfortable, drivers would take *any* road surface at the same high speeds, and driver and passenger safety became a major issue. Dirt roads, potholes, it didn't matter. The ride was smooth, so the tendency was to drive faster than was safe. The suspension had to be "detuned" so that it was not only comfortable, but also safe to use.

Similarly, Lotus built a vehicle that was controlled entirely by a joystick. Push forward to accelerate, pull back to brake, push left or right to steer. Again, the system worked just fine, but the test drivers hated it. When they wanted to slow the vehicle, they wanted to be able to push against something solid so they could feel in control. They were not comfortable with the idea of trusting their lives to a flimsy joystick. They preferred mechanical linkages and real tactile feedback.

The point is that it doesn't matter how new and innovative an interface is. If the user is not comfortable with it, it is a "bad" interface. The more comfortable the user, the "better" the interface.

Whether you prefer the Macintosh or Windows platform, there can be no doubt that everybody has truly benefited from the idea that applications should share the same interface. Once you have learned the basics of using one application, then you will always know where to go and what to do to perform any of the basic functions that are common to any application.

If you are developing for the Windows platform, then you know that the interface that you create has to be consistent with the "standard" interface. If it isn't, then your application will not find acceptance with the broad user public. It doesn't matter if it is the greatest application ever written. Users will find it cumbersome and unintuitive. It will make them feel clumsy, and they will avoid it. Once a workable interface has been established, people will feel uncomfortable with anything that does not conform to it.

Most users are quite obliging and will go through the most tortuous procedures for you. But only once. If the process is too painful, they just won't do it again, or they may even abandon the process in the middle. This is called "voting with your feet." If consumers don't like your application, they will simply walk away and not come back.

If this kind of thing happens with your applications, you will gain fame as a developer, but not the kind of fame that you will want to brag about. Usability is important for any user device, but this is especially true for WAP devices, where the inherent physical constraints make it harder than usual for the user to interact with.

Users already tend to get "lost" in applications, but this is especially true of a WAP application. It can be hard to provide all the data that users want, as well as all of the necessary visual cues to let them know exactly where in the application they are, where they have come from, what steps they have completed so far, and what the next step might be.

On top of that, users of mobile devices are often likely to be distracted by something happening in the external environment, and this makes the developer's job tougher.

WAP is a totally different medium from a normal computer interface, and new rules need to be developed to help the user get from point A to point B without becoming frustrated, confused, or overwhelmed. Phone.com in the United States has been established for some time, and they have an extensive knowledge base gained from intensive usability testing. Here is a list of the basic constraints of any mobile device for you to think about, so that you can create an interface that the mobile consumer will readily accept. As well as going over each of the interface constraints of WAP devices, and the ways that we can overcome some of the problems, I will also go over some of the things that can and should be done to help the user.

LOW BANDWIDTH

Time is money, and in the case of WAP, this is literally true. The user is paying for the time spent on your application. You shouldn't ever forget this fact, because the user certainly won't. (This, of course, comes down to the perceived value of the application. If the user gets excellent perceived value for the time and effort spent online, then they will be happy. However, if it takes ten minutes to get a weather report that they could have gotten for free by turning on the radio, they will not be happy.)

Even when the technology advances to the point where connection time is not an issue, and only the amount of data transferred is paid for (when GPRS technology comes online), there will still be people using WAP devices that use the cheaper low-bandwidth solutions currently in use. You will find yourself developing low-bandwidth applications for some time to come.

One point that separates the professionals from the amateurs is the naming of files and directories in URLs. Large URLs embedded in your code, like http://www.webdesignsinteractive.co.uk/wapforprofessionals/chapterone/introduction/default.wml, simply increase download time and fill the cache that much more quickly. As the user won't see the URL anyway, it would make the application more usable if you named the files and directories with as few characters as possible (while still maintaining readability, of course).

SMALL SCREEN SIZE

The small screen size is the probably the most obvious limitation of a WAP phone. You should never develop your applications for use on anything larger than a display averaging 4 lines by 12 characters. Larger displays obviously will exist in the future, and some do now, but the majority of phones will have small displays. Applications developed for smaller displays work fine on larger displays, but the reverse is definitely not true.

As well as making your data readable, you have to make your menus navigable. If you have a menu that is 20 items long, and the most lines you can get on the screen is four, users will have to click the down scroll button 16 times to reach the last item. I don't have to tell you that the user is not going to do this many times before giving up on the application altogether.

Therefore, you should try to keep your menu sizes as small as possible, and use specific titles wherever possible. If you find that you have some general categories, you should put them at the end of the menu.

To illustrate, let's say you are a user looking for the local weather report. You are scrolling down the main menu on your WAP phone, and you reach an item called "Information" before you come to anything called "Weather."

What do you do? The weather report is "information," isn't it? So you click to select the Information option only to discover that it is general information about the service provider. Not what you wanted at all, so you click to go back to the main menu, click down past the "Information" option, down another three items, to "Weather Report," finally.

As you can see, this example applies to many different areas. News, stock prices, and even help on using the application itself can all be classified under the general heading "Information," so you should try to use specific menu options that are unambiguous.

Remember, you cannot make it too simple. The user will not feel like an idiot if you make it simple, but they may if you make it too complex. Nobody likes to be made to feel like an idiot, so keep it simple. Users will appreciate you not assuming that they know exactly what you had in mind when you created a "Miscellaneous" menu option.

Your job as a developer is to get into the user's head and understand the user's requirements. If users have to learn *your* logic in order to use the application, then they won't feel comfortable using the application and will tend to not return to it.

TEXT ENTRY

If you have already tried to enter data into a mobile-device keypad, then you will know that entering text is definitely not a pleasurable activity. Although there are millions of SMS text messages sent every day, sending long e-mail messages from a mobile device can be difficult and also expensive since you are typically online the whole time. Although PDA's don't have quite the limitations of a mobile phone when entering text, it's still a lot easier to use a keyboard.

In addition, sending the special symbols that are required for a URL or e-mail address requires the user to learn where these symbols are on a mobile phone keypad. This is not obvious in many cases. For example, on the Siemens phone I use, I have to press the star (*) button three times if I want a forward slash (/) for a URL. For an ordinary period or full stop (.), I have to press the zero (0) button four times, and for the at sign (@), I have to press the number or pound (#) button three times.

This is *not* intuitive and must have been planned by some kind of sadist. So, if you have to prompt the user for text input, don't ask for anything complicated, and try to help the user wherever you can. For example, if you want the user to enter an email address, have *two* text boxes, one for the part before the "@" sign, and one for the second part of the email address, so that the user does not have to try and locate the "@" symbol at all. It is a trivial task for you as a developer to concatenate a couple of strings, but the total user hours that you will save altogether will be huge. And the users will appreciate the time and effort that you took to make their lives easier.

Number of Keystrokes

In terms of usability, remember to organize your application so that the user can get to the core of your application either immediately or with very few keystrokes.

Password Text Entry

One perfect example of designing for the wrong platform is the infamous password entry. In HTML, the common convention is to have an `<INPUT TYPE="PASSWORD">` tag, which displays an asterisk (*) for each character that the user types in. This is perfectly good security, and it works very well—on a PC.

On a WAP device, this convention makes it extremely difficult for the user to know what they are entering, because it often takes several button presses to reach the letter they want. If they can't get an instant visual check on what they have entered, it is all too easy to make a mistake and not find out until the application rejects their entry. Then, of course, the application returns them to the password-entry screen again. You only have to go through this once or twice yourself to understand how frustrating this can be to any user.

Software developers are creatures of habit, however. We all get into certain modes of thinking about something, and then we can't see any other solutions. With passwords, we cannot display the characters as the user types them, of course, because anyone else could read them over the user's shoulder, right?

Stop. Think again.

That thinking is fine for a traditional computing environment, but it does not mean that this same habit must be applied to all other uses. As far as WAP is concerned, it is not the best solution. If WAP users want privacy to enter a password, all they have to do is turn the screen away from other people. Just try, as an experiment, to covertly read some text from the screen of a WAP device when the owner doesn't want you to see it. Hopefully this will convince you.

You might be wondering whether this password data will be displayed again if the user presses the Back key. Well, that is not a problem for a developer—there are simple ways to handle this, and we will cover them when we get to Chapter 6.

Application Personalization

Once you have received some information from a user, you should keep it so that the user does not have to provide it again—ever. This applies to anything the user has ever entered into your application. If she has answered "Chinese" to your question asking her what kind of restaurant she is looking for, then the next time she is looking for a restaurant, put Chinese at the top of the list of available options.

Personalization also applies to logging into the application. If you have an application that requires the user to log in to use it, then you should be able to keep the subscriber ID or the phone number in the database, along with the password. Once you see that subscriber reappear, all you need to do is welcome him back with a message like "Hi Joe, please enter your password" so he does not have to enter his name again.

Aside from making the user feel important enough that even the computers know who he or she is, it means that the user can avoid having to enter data again and again. Nobody minds being asked a question once. Answering the same question over and over is irritating.

Data Field Entry

If data entry is painful, then any registration forms or surveys via the phone should be avoided. It is practically impossible for users to fill out a form on a mobile phone. And as mentioned earlier, it is extremely hard to enter e-mail addresses and URLs.

The "smart text" input found on some phones doesn't help the average user at all. It is good for "power users," but that is a very small minority of phone users. The rest of us still have to hunt and peck. That means that if data entry is needed, the number of fields that are to be entered via the phone should be kept to an absolute minimum.

NOTE: "Smart Text" is where the device tries to intelligently complete words that you have started to enter. If the correct word is selected after you have entered the first few letters, pressing a specific key will cause it to be accepted and the device completes the word and adds a space ready for the next word to be started. The trouble is that instead of searching for the letters on the keypad, you have to search the keypad and watch the display at the same time. Otherwise you can end up with some very strange stuff on the display where the device has completed words for you that are not always what you intended. Trying to go back to correct it can just make things worse. Unfortunately, it's an example of a very clever idea, that must have taken an awful lot of time and effort to get done, that doesn't actually work for the average user.

Users may go through a registration process if the application first gives the user enough perceived value, or incentive, to register, such as giving free stock quotes or other data with a perceived value. However, we have probably all been asked to register on at least one web site where we are promised access to special deals or privileged data, only to find that the actual value is negligible, or even negative if you find yourself on an "opt in" e-mail list that you can't get off. One or two experiences like this, and users learn to stay away from registering anywhere else.

Another way to give users some perceived value is by allowing them to window shop first, and only requiring them to register if they decide to buy. It will be obvious to the user that he has to give some personal data in order to receive the goods, and now there is a valid reason for being asked for the data.

If you are offering a type of application that can provide a repeat service, then you could allow users to go through the full registration process on a web site, and only ask the most basic questions via the phone keypad in order to authenticate them as valid users, and to take the details of their current request.

To summarize, mobile device keypads were designed primarily for making phone calls, not for data entry. Keep the data-entry requirements to a minimum.

USING THE CACHE

Using the cache properly can greatly improve performance, which is key to the perception of the user.

You must keep in mind that connection time has a cost to the user. On circuit-data networks, like most current implementations, the user will pay for the connection to your application just like they currently pay for a phone call, getting charged by the minute. When packet technology (GPRS) becomes widely used, the user will be charged a fixed fee for each data packet.

Whichever type of network is being used, you should try to construct your application so that it will work with as few round trips to the web server as possible, thereby minimizing the cost to the user. If the user knows that it is cheap, he will come back more often and also recommend the application to others.

So how do we minimize the number of round trips to the web server and get a happy user? We use the cache. The micro-browser in the phone has a memory, or RAM, cache for the data that has been downloaded from the network, just as browsers on PCs cache images when web pages are downloaded to the PC. Unlike the browser in a PC, however, you as a developer can specify exactly how long that data in the micro-browser's cache can be considered valid before it is overwritten.

For example, let's say you request the local weather report, but you are distracted by something else and disconnect after it has been downloaded but before you have read it. Because the data is in the cache, and the weather report is only updated by the service once every 12 hours, you can go through the same menus later and read the weather report without having to connect to the service to download the data again.

On the other hand, information like stock quotes should be flagged as data that can, and should, be overwritten either immediately or after just a few minutes. Each type of data has its own "volatility factor," and you should bear this in mind when downloading data to the user.

With the sensible use of this caching facility, you can allow the user to access the cache for a lot of his or her daily needs, such as menu structures, which of course makes them extremely fast to access, and then do the round trip to the service only for the final up-to-date data that the user is trying to reach. This means that if the user wants the latest weather or travel reports, she can navigate the menus all the way to the actual report, and the phone only contacts the service provider to get that report. The user is then only paying to download the data that she really wants.

Of course, the cache is not inexhaustible, so if the user is accessing a lot of different services, one after the other, then at some point the latest sites accessed will replace the oldest data in the cache. At that point, the menus would again be downloaded when the user returns to that site.

A WML deck is always downloaded as a complete item. You can't just update one card in a deck from the server. If a particular card calls for information from the server, the entire deck containing the needed card is also downloaded, and consequently cached as a complete item. This means that you can actually plan out a "caching strategy" for your WML cards and decks so that when the user clicks on an option that calls up an often-used submenu, you can ensure that the deck is kept in the cache from the first download, and the needed card will appear immediately. This is a simple technique, but one that can make the application appear to be lightning fast to the user, which is a very good thing.

TYPES OF WML CARDS

There are three basic types of cards that you will construct in WML.

- ▼ **Choice card**   Generally presents a set of menu options to the user.
- ■ **Data entry card**   Requests data from the user for use within the application.
- ▲ **Display card**   Displays or presents the result data to the user.

Each of these card types have their own basic rules as to how they should or shouldn't be constructed, so as to keep the user feeling as "comfortable" as possible.

In this discussion, we'll also come to the concept of "soft" keys. Most WAP enabled mobile phones, for example, have one or two unlabelled buttons, or keys, that are usually placed under the LCD screen. These keys can be re-programmed "on the fly" by you as the developer via the application. The labels for these keys normally appear on the bottom of the LCD screen above the keys, and are usually labels like "Back" and "Options". You can reassign the labels that appear on the screen and/or the actions that are taken when one of the keys is physically pressed, on a card-by-card or deck-by-deck basis. PDAs or devices with touch sensitive screens may well have these keys as onscreen buttons that you can press directly rather than a physical key.

Choice Cards

A choice card presents a series of choices to the user, and it should have a short title of 12 characters or less, which will appear at the top of the display. If the title wraps onto the next line, you will have two things occurring. Firstly, the whole screen appears to "jump" up and down one line as the user moves on to the next card. Secondly, you will lose one usable line from the display.

If the title, when originally worked out, is a long one, such as "Traffic Reports for London," simply modify it so that it fits into 12 characters. In this case it could become "Traffic News".

Keep the number of choices to as few as logically possible. If there are more than will fit onto the screen, it will force the user to scroll down to see them all. Also, the Phone.com browser allows you to assign a number key (1 to 9) to each menu item for the user to press, instead of first requiring the user to scroll down and then "select" the option, which means you should have no more than nine items on one menu to ensure compatibility.

For the choice cards themselves, there are several different types: Data lists, Navigation only, Action/navigation, Pick option, and Change option.

Data Lists

Data lists are used for things like names, addresses, restaurants in an area, and that kind of thing. Any list of data items based on a general selection by the user will probably be of this type. From the list, the user selects one and is typically taken to an in-depth view of that data item.

The data should always be displayed in a sorted order and should be sorted by the most appropriate column. For a list of names, the list could be displayed alphabetically by surname, first name, and so on. A list of restaurants could be sorted by restaurant name, food type, and so on.

The first soft key, or the one most likely to be used as a "Select" option, should be used to drill down further into the data. So something labeled "Select" or "View" would normally be used for this.

The second soft key should be programmatically assigned an action that you might want to perform on the selected item, like "Edit", "Dial", or "Done" (to go back), for example, or it could lead to a submenu that gives a choice of actions to be taken (a typical example would be "Add", to add a new item, "Edit", or "Delete").

Navigation Only

Navigation-Only choice cards are straight navigational menu choices. These cards should be ordered by the most-often used choices first, or just alphabetically if it is a random list. This gives the user some reference points to go by.

The first soft key should be labeled "Select" or "OK" so that the user can select the option or go on to the next submenu. The second soft key should either be labeled "Menu", which leads to a menu of other navigational options, or should just be left blank.

Action/Navigation

An Action/Navigation card is used either to perform an action on an item, such as dialing the selected name in list of addresses, or to go back to the main list from an already selected name in the address list. These cards should be ordered with the most often used choices first.

The first soft key should be labeled "Select" or "OK" so that the user can either select the option or see more. The second soft key should either be labeled "Menu", which leads to a menu of other navigational options, or labeled as the action item required, such as "Edit", or "Done", to go back.

It is important to note that you shouldn't mix items of different types, such as having a "Delete" option on a menu of navigation options, or a "Back" option on a list of action items. This will only serve to confuse the user.

Pick Option

The Pick option is used for picking an option from a selection, such as choosing colors, a flavor of ice cream, a location, and so on. The options should be sorted by whatever makes sense, which in most cases will probably be alphabetically by name.

The first soft key should be labeled "Select" or "OK", so that the user can select the option or see more. The second soft key should be labeled "Cancel", which will take the user back to the previous menu or deck, knowing that the selection was actually cancelled. Please note that this *not* the same as "Back".

You should also if possible pre-select the common or default item so as to make life easier for the user. If you are going to allow user input, then the normal convention is to put an item at the end of the list called "Other" that will take the user to a free entry text box.

Change Option

The Change option is used for changing an option of some kind, such as a parameter that has already been set or input. If there is a list, then it should be sorted by whatever characteristic makes most sense and gives the user the maximum usability.

The first soft key should be Save. The second soft key should be Cancel, which will take the user back to the previous menu or deck, knowing that the selection was indeed cancelled (this is not the same as Back).

Entry Cards

When you are collecting data from a user, the entry cards will contain the input fields that you collect the data with. An entry card should have a short title of 12 characters or fewer. (We already covered the reasons for this in the "Choice Cards" section.)

You should use the format attribute of the `<input/>` tag wherever possible (see Chapter 6), to minimize any effort required by the user, and to ensure maximum readability for the user. You should avoid making the user enter text wherever possible, and use choice lists via the `<select/>` tag so that the user only has to scroll and press the "Select" key rather than type anything.

With the continued aim of making life easier for the user, you should use the minimum number of input fields possible, and of those that you do use, make as many as possible optional for the user.

An entry card is similar in functionality to a data entry form on a web browser. If you do have a lot of data to be collected from the user, you should break it up into two or more decks, with "Send" or "Submit" key presses required at the end of each one. Aside from preventing catastrophic loss of data, this also gives the user a little more space. Imagine entering what seems to be your entire life history in one marathon half-hour session, only to be told something like "Connection Lost," just as you were about to press the "Send" key, and consequently discover that all that time and effort has been wasted. This is not even vaguely amusing from the user's standpoint, I assure you.

Entry cards should have one and only one soft key, which should be labeled "OK" (if no more data is needed on a single entry card), "Next" (if more data is needed, or you are taking the user on to a confirmation card), or "Save", "Done", "Send", "Cancel" or something similar (if you are at the end of a form). This avoids all confusion.

Finally, password entry has already been covered, but this is really worth repeating. *Don't* hide the user input from the user. This is not a PC—all users have to do to hide what they are entering is to turn the phone away from other people.

Display Cards

A Display card is one that displays or presents the result data to the user. As this is what the user is using the device and paying for, it is fairly critical that the results can be read easily and without confusion. It is also important that the user cannot accidentally hit a key and inadvertently lose the data that he or she may have spent some time reaching.

You should avoid using or displaying words longer than ten characters, if at all possible. Words that are forced to wrap to the next line on the display are much harder to read than words on one line. Although it is not safe for mobile device users to use them while driving, this will happen whether laws are passed against it or not. All we can do as developers is to try to make life less dangerous for the users. Words that wrap will take the user's attention, and if they are driving, that momentary lapse could prove fatal. It would look pretty strange on the coroner's report if the user died "while trying to read the word *sarcophagus*."

You should also display titles as needed to always provide the correct context to the data. For example, stock and share information should have the name of the stock at the top of the screen.

If no other actions are allowed on the data, you should use the first soft key for "More". Or, if it is used for navigation, the first soft key should take the user to the next most logical place the user will want to go. If the next most-common place to go is to the next item in a list, such as to the next e-mail message or the next stock report, then make the soft key "Next". If the most common place to go is back to the list to get the next item, then make the soft key "OK", and then take the user back. You should never put actions like "Delete" on the first soft key, when users have been using it for "OK" or "Next". This is just a disaster waiting to happen.

The second soft key should be an action item, if needed, such as "Delete", "Reply", or "Menu", which could lead to a submenu of actions or navigational options.

Speaking of disasters, if you do ask the user a question as part of the display process, such as "Delete?", then you should always make the first soft key the *least* dangerous option, such as "No" in this example. This way, if the user presses the first soft key without thinking, because they really wanted to go to the next item, then all they can lose is a little time and not that crucial e-mail with the latest proposal for that big contract.

THE "BACK" BUTTON

When using a mobile phone, users do not feel that they are going "down" through a hierarchy of menus. Instead, users have the sense that they are traveling "along" through the application.

If you add to this the fact that there is no clearly defined "Home" button on a mobile phone, then it is no surprise that users will press the "Back" button as many times as is necessary to return to where they want to be in the application.

In actual fact, the "Home" function is often carried out by the mobile phone's "Clear" key being held down for a second or so. Even knowing this, I personally prefer to use the "Back" key. Like many other phones on the market, my "Clear" key also doubles as the "On/Off" key, so if I hold the "Clear" key down for a fraction of a second too long, I don't go "Home", I go Goodbye!

However, there are times when you don't want anyone to be able to just go back through the sequence of screens that they have just come through. If you are already logged into the application, and one of the screens was a password screen, then it would be unnecessary to go back, as well as not very good security if you have followed my advice about not using asterisks for passwords.

There are ways of channeling the user through the application by defining what action the card takes, depending on whether the user has entered the card coming "forwards" through the application or "backwards" from a later screen. This is covered in Chapter 6. This is a very efficient way of handing the history stack, and it could be put to very good use in a normal web site if the syntax were available in HTML.

GRAPHICS

Because of the constraints of the WAP devices themselves, the subject of graphic display is a bit "hit-and-miss" at the best of times. To begin with, you cannot assume that the devices your application will be sending to can even support graphics in the first place. This is one of the reasons why the ALT attribute has been made mandatory: to ensure that the user does actually get to see something of the intended message.

Where the graphics are accepted, you must keep the sizes down to fit the smallest display; otherwise the results will definitely not be as you had intended.

Some micro-browsers, notably the Phone.com micro-browsers, have a set of graphic icons actually built into the micro-browser itself. This can make life easier for you, the developer, if you want to display an e-mail icon, or a send icon, or whatever. You simply tell the micro-browser that you want icon X displayed. Job done. While this will only work on Phone.com micro-browsers, it has the advantage that these commands are simply ignored by other micro-browsers, but can look really nice on a Phone.com menu. The other major advantage of this approach, of course, is that these icons don't have to be downloaded, and so the application can run faster.

CHAPTER 4

WAP Development Tools and Software

Before we get into the syntax of WML and how it all hangs together, let me ask you a fairly obvious question. If we all know HTML, why not just use HTML for all WAP devices instead of creating a brand-new syntax with its own rules? Haven't we all got enough to do without getting into a whole new learning curve? The answer to this is largely irrelevant, as WML is what we have now, but here are some points to consider:

- ▼ HTML is not actually very robust as a language. Even though it has been around for several years, there are still major differences in behavior between Internet Explorer and Netscape when it comes to displaying the carefully formatted HTML. This has been a thorn in the side of web developers for a long time. (Personally, I believe that Netscape lost the plot some time ago, but hey, what do I know? I've just been one of those web developers cursing because I omitted a <form> tag, and none of the labels that I filled dynamically from a carefully designed database showed up in Netscape, although Explorer understood perfectly what I was trying to do and displayed them all beautifully.)

- ■ HTML is an interpreted language and is transmitted as straight ASCII code to be formatted by the browser. WML is binary encoded in order to reduce the amount of data that must be transmitted between the handset and the base station as much as possible. The current bandwidth is only 9,600 bps. The encoding is done transparently at both ends of the connection and does not impose itself on the user. This does, however, demand that the developer follow the rules of WML very strictly. If the rules are not followed exactly, the code will not work at all; whereas HTML will at least partially work and display an error.

- ▲ The micro-browser does not have to deal with a lot of formatting because of the screen limitations. So if you are going to start tearing HTML to pieces, you may as well build it from scratch to include some real enhancements based on hard-won experience.

The basic enhancements to WML that have been added to make it a different beastie from HTML are these:

- ▼ WML contains a built-in event model. HTML does not. HTML requires JavaScript or VBScript, or ActiveX components to provide event handling. Although WMLScript is supported by WML, it is not required in devices that do not need to provide call handling or phone access functions.

- ■ The WAP standard includes a call-handling model (WTA), which HTML does not include and never will.

- ■ The WAP standard includes (or will include in future versions) telephone device interfaces (WTAI), such as phone/address book access, and SIM card access.

- ■ WAP was specifically designed to minimize bandwidth usage on slow or low-bandwidth carriers (via binary encoding, as already mentioned). HTTP is "optimized" for TCP/IP and socket communication via ASCII text, and any nonoptimum code included (for example, a row of 2,000 space characters) is transmitted without inspection.

▲ WAP includes support for encryption (WTLS), even when the device does not support smart cards (which is where encryption normally lies).

So, if you want to develop new WAP applications that are going to take the world by storm, the first thing to do is to select one or more WML editors to write the code in. More importantly still, you need to pick the emulators that you are going to use to test the written code. While not guaranteed to exactly emulate the target micro-browsers, they are far better than not testing at all.

EDITORS AND EMULATORS

Even though WAP development is still very much in its early days, there are already a number of software packages and development environments available for free download. Each of these has its wrinkles, of course, and most are platform specific. But that's OK and, in fact, to be expected, until the day I take over as Software Coordination Manager, Planet Earth. Until that day (still a little ways off), what I do personally is to keep all of my code in a "common" directory, and have all of the editors and emulators point to that directory as a default. This way, when I have written my code in any editor, I can quickly test the same code in the other emulators to make sure that the code I have written is actually platform independent.

There is no substitute for testing on the actual devices, but until you can afford a few hundred different models of phones—with more coming on the market all the time—the emulators will have to do. In fact, if you keep within the guidelines as laid out in the previous chapters, you can actually get away with emulating any phone that has a Nokia micro-browser installed and any phone that has a Phone.com micro-browser installed. The remaining emulators can be used to see how the applications or sites will look on a specific model of phone, but at least the basics will be in for any phone.

The following sections provide a list of editors and emulators that are available to download and use. Please bear in mind that this list is incomplete even as it is being written, as new products are being released all the time. But at least it should get you started.

WAP Editors

WML Express

Produced by	Astrosolutions
Download from	http://www.muenster.de/~sak/wml.htm
Platforms	Java

This WAP word processor is available in online and downloadable versions, but currently it is all in German. Unless you know German and can understand the instructions, don't bother.

WAPTor

Produced by	WAPTop, s.r.o.
Download from	http://www.waptop.net/default.htm
Platforms	Windows 95/98, Windows NT, and Windows 2000

WAPTor is a WML editor for Windows 95/98, NT, and 2000 systems. With it you can open and edit existing WML files, or create new files from a simple template. I have this open on my desktop all the time. It is great for writing small chunks of code on the fly.

WML Editor

Produced by	Jan Winkler
Download from	http://www.jan-winkler.de/dev/e_wmle.htm
Platforms	Windows 95 or above

This is another offering from Germany, which has some nice features. For example, you can access all of the Library functions from a tabbed window. You have to switch to English as a language by selecting "Program/Sprache/English" from the menu bar, otherwise you have to speak German.

Textpad

Produced by	Helios Software Solutions
Download from	http://www.textpad.com
Platforms	Windows 9x, NT, 2000

Textpad is a simple but powerful replacement for Notepad. You can use it as a tool for editing your web pages, or a programming IDE. You can plug in "clip libraries" for just about anything, from HTML tags, to JavaScript, to Perl. Now there is a WML clip library add-on available for Textpad as well, and it is fairly comprehensive.

Nokia WML Studio for Dreamweaver

Download from	http://exchange.macromedia.com
Platforms	Windows

Nokia WML Studio for Dreamweaver is a downloadable extension to Dreamweaver that enables users to create content for Wireless Application Protocol handsets. The extension launches from within the Dreamweaver HTML editing environment and provides a What You See Is What You Get development environment for the Wireless Markup Language. The extension includes a WML 1.1 parser with visual error feedback and a preview function that can display WML content in multiple Nokia WAP mobile phone interfaces, running in a conventional Internet browser. That's the official line. If you own Dreamweaver, then you should try it for yourself. Personally, I don't use it.

ScriptBuilder 3.0 WML Extension

Produced by	Netobjects
Download from	http://www.netobjects.com/products/html/nsb3wml.html
Platforms	Windows

This extension kit supplies WML support for the ScriptBuilder 3.0. I haven't tried this, as I don't use ScriptBuilder.

Dot WAP 2.0

Produced by	Inetis Ltd
Download from	http://www.inetis.com/english/solutions_dotwap.htm
Platforms	Windows 95 and higher

Dot WAP is a visual tool for WAP site construction.

XML Writer

Produced by	Wattle Software
Download from	http://xmlwriter.net/
Platforms	Windows 9*x*/2000/NT4

This editor provides users with an extensive range of XML functionality such as validation of XML documents against a DTD or XML Schema, and the ability to convert XML to HTML using XSL style sheets. Users can also combine CSS with XML for direct formatting of XML data.

MobileJAG

Produced by	WAPJAG
Download from	http://mobilejag.com
Platforms	None. Runs from any web browser.

MobileJAG is a free online service, and there's no need for downloading and installing software. Everything's online. You have to register with them to get a login in order to use the software and create yourself a simple WAP site.

WAP Emulators

Yospace

Produced by	Yospace
Web site	http://www.yospace.com

The SmartPhone Emulator is impressive. Although it is not free, you can set up a workspace environment that allows you to view your application in a number of handsets simultaneously. It is available for Windows, MacOS and UNIX-based platforms.

EzWAP 1.0

Produced by	EZOS
Web site	http://www.ezos.com

EZOS provides EzWAP, the first platform-independent WAP browser enabling all kinds of computing systems to access the mobile Internet: mobile devices (PDA, Pocket PC, PC Companions), mobile computing and embedded systems, PCs running Microsoft Windows NT, 2000, CE, and so on.

WAPalizer

Produced by	Gelon
Web site	http://www.gelon.net

The script fetches WML pages from WAP sites and converts them to HTML on the fly. This means that you will be able to view most WAP pages, but some pages, especially those with a lot of input forms, are very difficult to convert to HTML.

Ericsson R380 Emulator

Produced by	Ericsson
Web site	http://www.ericsson.com/developerszone

The R380 WAP emulator is intended for testing WML applications developed for the WAP browser in the Ericsson Smartphone R380. The emulator contains the WAP Browser and WAP Settings functionality that can be found in the R380. The R380 WAP emulator can be downloaded from Symbian.

NOTE: Membership in Ericsson Developer Zone is required.

WinWAP

Produced by   Slob Trot Software Oy Ab
Web site   http://www.slobtrot.com/winwap/

WinWAP is a WML browser that works on any computer with 32-bit Windows installed (Win95, Win98, WinNT). You can browse WML files locally from your hard drive, or from the Internet with the HTTP protocol (same as your normal web browser).

WAPMan for Windows 95/98/NT

Produced by   wap.com.sg
Web site   http://www.wap.com.sg/downloads/

The WAPMan is a portable browsing device, combining access to the Internet with the properties of a mobile phone. With its unique WAP gateway structure, the WAPMan has fast downloading capability and is highly compact and portable, functioning as a mobile commerce and lifestyle portal.

WAPsody

Produced by   IBM
Web site   http://alphaworks.ibm.com/aw.nsf/techmain/wapsody

WAPsody simulates most aspects of WAP. It is designed for use as a WAP application building environment. Unique to WAPsody is its ability to reproduce the behavior of the underlying network bearer service and the protocol layers that build on it. This feature can be used to simulate the exact behavior of a WAP application both in terms of usability and communications efficiency.

The WAPsody simulation environment can be executed stand-alone, or can execute demos that are being hosted on the WAP infrastructure at IBM Zurich Research Laboratory.

WAPsilon

Produced by   Wappy
Web site   http://wappy.to

WAPsilon converts WAP sites to HTML that can be viewed on the Wappy site or on their "device." WAPsilon can be integrated into your web site or plugged into your browser.

Opera

Produced by   Opera
Web site   http://www.opera.no

Opera is actually an HTML browser that now supports WML.

WAPEmulator

Produced by	WAPMore
Web site	http://www.wapmore.com

There is a new Netscape version that is not as functional as the Internet Explorer version.

Wireless Companion

Produced by	YOURWAP
Web site	http://www.yourwap.com

With the Wireless Companion you may access any WAP and web content over the Internet, including the free personal wireless services at yourwap.com.

Pyweb Deck-It

Produced by	Pyweb
Web site	http://www.pyweb.com/

This is a WAP phone emulator that supports the Nokia 7110. More models are planned to be added as they are completed. With this program you can browse WML sites on the Internet.

M3Gate

Produced by	Mobile Media Mode
Web site	http://www.m3gate.com/

WML and WMLScript are fully supported. The appearance and user interface of M3Gate may be designed according to the wishes of the customer. M3Gate gets stream data using Internet Explorer or Netscape Navigator as a transport.

SOFTWARE DEVELOPER KITS (SDKS) AND INTEGRATED DEVELOPMENT ENVIRONMENTS (IDES)

Table 4-1 shows each of the SDKs and IDEs available at the moment. All of these have their advantages and disadvantages, useful features, unintentional features (bugs), and different ways of creating the WML files and testing them. The only two of these that I would consider to be mandatory are the Nokia WAP Toolkit and the Phone.com SDK. If you do absolutely nothing else, you *have* to see what your code is going to look like in these two emulators. These are two completely different micro-browsers, with totally different display characteristics. Although the predominant browser in the United States is the Phone.com version, you can only ignore one or the other at your own peril.

Chapter 4: WAP Development Tools and Software

SDK Provider	SDK	URL	Emulator	Editor	Debugger	Samples	Help
ThinAirApps	ThinAir Wireless SDK	http://www.thinairapps.com	Yes	Yes	Yes	Yes	Yes
MobileDev	MobileDev	http://mobiledev.speedware.com/		Yes	Yes	Yes	Yes
Wapalize	Wapalize WAP Development Tool Kit	http://www.wapalize.co.uk/	Yes	Yes		Yes	Yes
Nokia	Nokia WAP Toolkit	http://forum.nokia.com/main.html	Yes	Yes	Yes	Yes	Yes
Motorola	Mobile Application Development Kit (ADK)	http://www.motorola.com/MIMS/MSPG/cgi-bin/spn_madk.cgi	Yes	Yes	Yes	Yes	Yes
Ericsson	WAPIDE SDK 2.1	http://www.ericsson.com/developerszone/	Yes	Yes			Yes
Phone.com	UP.SDK 4.0	http://www.phone.com	Yes			Yes	Yes
WAPObjects	WAPObjects	http://www.wapobjects.com/wapobjects/en/		Yes	Yes	Yes	Yes
WAPMine	WAPPage 1.0	http://www.wapmine.com/Products.asp		Yes		Yes	Yes
Perfect Solutions	CardONE	http://www.peso.de/wap_en/index.htm	Yes	Yes			
Dynamical Systems Research	WAP Developer Toolkit 1.0	http://www.dynamical.com/wap/index.html	Yes	Yes	Yes	Yes	Yes
PWOT	PWOT WML-Tools	http://pwot.co.uk/wml/	Yes		Yes	Yes	Yes

Table 4-1. Developer Solutions - Integrated Development Environments

CONVERTING IMAGES

There are tools that don't really fit into either of the above categories, but are extremely useful nevertheless. One of these is Pic2wbmp. As the name suggests, it is a software tool for converting images from PhotoShop (PSD files) into WBMP (wireless bitmap image) format.

The WBMP format is the only image format that has been specified by the WAP Forum. Some SDKs support GIF or BMP formats, but these are not guaranteed to work in a micro-browser. On the contrary, these formats are currently a liability for a WAP device because of the image sizes and download bandwidth requirements.

The WBMP format supports the definition of compact image formats suitable for encoding a wide variety of image formats and provides the means for optimization steps such as stripping of superfluous headers and special-purpose compression schemes. This leads to efficient communication to and from the client and efficient presentation in the client display.

A WBMP image has the following characteristics:

- Compact binary encoding
- Scalability, which means future support for all image qualities and types (color depths, animations, stream data, etc.)
- Extensibility (unlimited type definition space)
- Optimization for low computational costs in the client

Specification of Well-Defined WBMP Types

NOTE: This is for techies only. If you don't want to get glazed eyes, skip to the summary of this chapter.

WBMP Type 0: B/W, Uncompressed Bitmap. WBMP type 0 has the following characteristics:

- No compression.
- Color: one bit with white=1, black=0.
- Depth: 1 bit deep (monochrome).
- The high bit of each byte is the leftmost pixel of the byte.
- The first row in the data is the upper row of the image.

The WBMP image data is organized in pixel rows, which are represented by a sequence of octets. One bit represents one pixel intensity with value white=1 and value black=0. In the situation where the row length is not divisible by 8, the encoding of the next row must start at the beginning of the next octet, and all unused bits must be set to zero. The data bits are encoded in a big-endian order (most significant bit first). The octets are also arranged in a big-endian order—the most significant octet is transmitted first. The most significant bit in a row represents the intensity of the leftmost pixel. The first row in the image data is the top row of the image. (Sources: www.wapforum.org, www.nokia.com)

SUMMARY

Armed with a selection of the above tools, developer's kits, and image conversion software, there is nothing that you can't create as far as a WAP application is concerned. We have the tools; now what are we supposed to do with them? Let's move on to the code itself.

CHAPTER 5

Working with WML

When you hear of a Web site being in *WAP format*, this means that a copy of the existing Web site's original HTML content has been converted to WML, which greatly reduces download times and simplifies the presentation for WAP devices. Unfortunately, this conversion is not a two-way street. If you were to display WML pages in a PC Web browser, it would look very boring indeed—you can confidently say that WML will never replace HTML. Modern versions of Microsoft Internet Explorer and Netscape both have integrated WML browsers now, so you can look at a WML page from your Web browser yourself and see what I mean.

What this means is that the WML, or "WAP enabled" site, has to exist independently of the HTML site. It is possible to dynamically return either WML or HTML to the browser if you create a page on the server with something like Microsoft Active Server Pages that can detect the calling device, and dynamically build either an HTML page or a WML deck depending on what it finds, but this kind of site would take a fair amount of resources to build and maintain.

Normally, however, the server will have a default "index" page that will be searched for first if there is an unqualified URL passed. (Which is the normal way that people enter a Web site address: as a domain name, without the actual file name.) So if the default page is "index.htm", then any request to the server will get this page returned. A micro-browser pointed at the same domain will also get the same page returned and it just won't work. Consequently, the default WML page, "index.wml" has to be in a separate directory so that the right page is returned to the right browser.

The current convention is to have a main Web site, and a directory off the main site called "/wml". So, for example, the main site would be

http://www.webdesigns.ltd.uk/

and the WML version of the same site would be

http://www.webdesigns.ltd.uk/wml/

WML BASICS

WML is a fairly basic markup language all by itself. The basic elements of WML can be learned in two hours or less if you give it your undivided attention, but as with most things, the real learning and practical experience comes after the introductory period when you find yourself having to solve specific problems for specific clients.

Like HTML, WML allows developers to specify the format and presentation of text content on pages, to control the navigational structure of the pages within the site, and to link pages to other pages that are either internal or external to the site. Because there are some real differences in the way that the pages are controlled and sent out to the client in comparison to HTML (there is a physical size limit to the page, and it is split into a discrete series of screen-sized "sub-pages"), the individual "sub-pages" are called *cards*, and with true developers' logic and sense of humor, a collection of cards is a *deck*.

WML, by itself, produces static "sub-pages", or cards. Static cards only display text, or allow users to jump from one card to another through links. Compared to the tools and effects the Web developers have at their fingertips, WML cards are not too exciting.

To make up for this, WML has its own version of JavaScript, called WMLScript. WMLScript allows developers to add functionality to their WML cards and decks. Currently, WML files and WMLScript files must be saved independently of each other, which means that a number of files may be associated with a single WML deck.

There are several conversion programs already on the market to convert existing HTML sites to WML sites. Some of these applications claim that they will transform am entire Web site to WML. This may sound like a great idea, but there has to be a judgment call on this issue. The WAP micro-browser is so different from a PC browser that the intent of the original HTML content may be seriously compromised if the conversion is left completely to a mechanical process. This is particularly true if the design guidelines in previous chapters are not adhered to. Knowing when and how to make such compromises will always be beyond the ability of any conversion software. The alternative is to create WAP content directly for mobile devices as a separate activity.

Of course, there are currently well over one billion Web pages on the Internet. Google, alone, currently claims to have over 1.3 billion pages indexed. I am sure that great many of these with suitable content for access by a mobile device could be mechanically converted—it would certainly be a lot of hard work to have to re-create most of these pages manually.

WAP and the Web

WAP is an open specification, and is comparable to many other protocols used in existing Internet technology. The major difference between WAP and any other set of protocols is that WAP is optimized for the restrictions of WAP devices.

WAP covers the *application level* (WML and WMLScript; collectively known as WAE, the Wireless Application Environment) and all of the underlying transport layers (WSP and WTP, WTLS, and WDP). The fact that the WAP Forum defines all of these protocols means that everything should interact without any problems.

The *bearer level* of WAP depends on the type of mobile network. It could be SMS (Short Message System), CSD (Circuit Switched Data), GPRS (General Packet Radio Service), CDMA (Code Division Multiple Access), or any of a large number of possible data carriers. Whichever bearer your target client is using, the development process remains the same—the other levels are completely transparent to both the developer and user alike.

Table 5-1 shows the layers used for both the Web and WAP, just to point out how they are similar. In actual fact, you will probably never have to refer to it again.

In WAP, you can consider WML to be the equivalent of HTML, which is used to describe the text content and the look of pages viewed in Web browsers. For graphical content in WAP, the monochromatic graphics a WAP micro-browser should be able to display are WBMP images.

HTML is used to encapsulate both the form *and* content of a page, and modern Web pages are often created as works of graphic art. Unfortunately, WML doesn't offer anything like that degree of control, and if you're expecting absolutely positioned pixel-perfect WAP pages, then you're going to be very disappointed.

	Web	WAP
Markup Language	HTML	WML
Scripting Languages	VBScript JavaScript	WMLScript
Session and Transaction Layers	HTTP	WSP - Wireless Session Layer WTP - Wireless Transaction Layer
Security	TLS-SSL	WTLS - Wireless Transport Layer Security
Delivery Service	TCP, UDP	WDP - Wireless Datagram Protocol
	IP	Bearer

Table 5-1. Web / WAP Protocol Layer Comparison

WAP devices and their characteristics are extremely varied—from tiny mobile phone screens to larger PDA displays—and WML makes a minimum number of assumptions about how the page will be displayed on the device's screen. This means that although you can be fairly sure that the WML page will contain the content you originally wanted to display, the appearance of the content will depend on the device.

You can, of course, write separate code for each device, but as there are already hundreds of devices with more coming onto the market all the time, testing pages across all possible client devices is impossible.

Having said all that, most Web designers will find that writing in WML is really quite straightforward. WML has been designed with the experience of HTML's evolution, and the end result is very workable.

Writing WML Code

We'll first look at the basics of WML so that you can pick up the foundations easily. Once you have gotten the basics, there is a mass of information out there on the Web that you can plug into to improve your expertise. In the next few pages, we will develop a simple WAP site. For the sake of originality and creativity, we will call it "Hello World."

Keep in mind that the display screen of a mobile phone is smaller than that of a PC. This means that most WAP sites and applications will actually be made up of a number of pages. We'll see in a moment that the Hello World site is made up of two pages: the first page displays a message to click the Accept button; the second page displays our "Hello World" message.

Chapter 5: Working with WML

In the world of WAP, *pages* are called *cards*. Thus, our Hello World application is actually made up of two cards. The total collection of these cards is called a deck.

Cards have a specific syntax that must be used

```
<card id="name"
<content>
</card>
```

The possible tags that you can use within a card are shown in Table 5-2. Obviously you don't need to use all of them, and they are shown here altogether as a reference for you to see the kind of things that we can do. Experienced HTML coders will immediately recognize the majority of them straight away. But in our "Hello World" site code following this, we only use the <do> and <p> tags to keep it simple for now.

Tag	Description
<onevent>	Associates a state transition, or intrinsic event, with a task.
<timer>	Provides a method for invoking a task automatically after a specified period of user inactivity.
<do>	Associates a task with an element within the user interface.
<a>	The short syntax for anchors, which can only be used to define (implied) <go> tasks that require a URL specification.
<fieldset>	Allows you to group multiple text or input items within a card.
<img>	Instructs the device to display an image within formatted text. Note that not all devices can display images.
<input>	Lets the user enter text, which the WML then assigns to a specified variable.
<select>	Specifies a list of options from which the user can choose. You can specify either single- or multiple-choice <select> elements.
<p>	Specifies a new paragraph and also has alignment and line-wrapping attributes.

Table 5-2. Tags Allowed Within a Card Element

THE "HELLO WORLD" EXAMPLE

There's nothing like jumping right in, so first we'll take a quick look at the "Hello World" WML code and what it looks like on a mobile phone's screen. The following sections will explain the various parts of the code and what they do.

Here's the code:

```
<?xml version="1.0"?>

<!DOCTYPE wml PUBLIC "-//WAPFORUM//DTD WML 1.1//EN"
"http://www.WAPforum.org/DTD/wml_1.1.xml">

<wml>
<card>
<do type="accept">
<go href="#hello"/>
</do>
<p>
Press the OK Button
</p>
</card>

<card id="hello">
<p>
Hello World!
</p>
</card>

</wml>
```

If you run the above code through an emulator, you will see the screen shown in Figure 5-1.
Then when you press the OK button, you'll see the screen in Figure 5-2.
Let's look at this code, one section at a time.

The Document Prologue

The first part of the code is as follows:

```
<?xml version="1.0"?>
<!DOCTYPE wml PUBLIC "-//WAPFORUM//DTD WML 1.1//EN"
"http://www.WAPforum.org/DTD/wml_1.1.xml">
```

This is known as the document prologue. For the most part, just think of this as a header that you must type at the beginning of every WML deck you create.

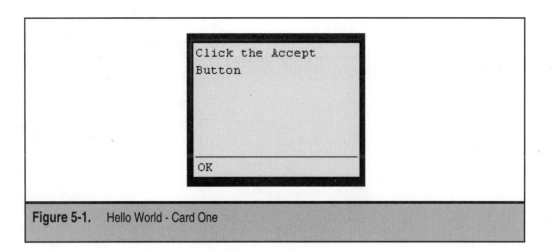

Figure 5-1. Hello World - Card One

The document prologue must contain the following:

▼ **An XML declaration** This specifies the XML version being used. XML is the language that WML is based on. This declaration is the first line of the preceding code.

▲ **A document type declaration (DTD)** This identifies the file type as WML and specifies the location of the DTD to use for compiling the file. The DTD is lines 2 and 3 of the preceding code.

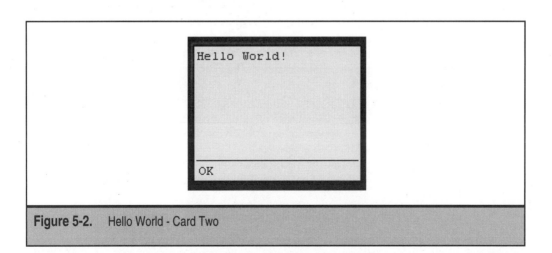

Figure 5-2. Hello World - Card Two

To keep this simple, all you have to know is that you must include these lines in every WML deck that you create, or it won't work. Just build a deck "template" file that contains these lines at the top so you don't have to keep typing them over and over. The only thing that would change in the future is the reference to the "wml_1.1.xml" file, as the specifications will change over time.

The Deck Header

The next line of the code reads as follows:

```
<wml>
```

The `<wml>` element designates the beginning of a deck. You can think of it as being the same as the `<HTML>` tag in HTML. Again, to keep it simple, the first line, after the document prologue, of every WML deck you create will be `<wml>`.

The First Card

The first card, or "page," of our application comes next.

```
<card">
<do type="accept>
<go href="#hello"/>
</do>
<p>
Press the OK Button
</p>
</card>
```

The first line here, `<card>`, designates the beginning of our first card. The next three lines use the `<do>` element to link an action (such as going to a particular URL) with a means of invoking it (in this case, pressing the "OK" soft key). The `<do>` element contains an attribute called `type`, which identifies the key that triggers the specified `<do>`. In this example, we have set the type as "accept". The `<go/>` element gives us the URL to go to when the Accept soft key is pushed. In this case, we will go to the "hello" card (the second card in the deck). The `</do>` line closes our `<do>` element.

The next three lines use the `<p>`, or paragraph, element to print the text "Press the OK Button" on the display screen.

The last line, `</card>`, closes the first card.

The Second Card

The second card, or page, of our application is defined with the following lines:

```
<card id="hello">
    <p>
        Hello World!
    </p>
</card>
```

In the first line, we've given this card an ID. The name of this card is "hello". (As you have already seen, the <go/> element in the first card references this card when the "OK" button is pressed.) Note that it is in all lower case to match the calling link exactly. If it was called "Hello", with an upper case "H", then you would get nothing happening when you pressed the "OK" soft key.

The next three lines use the <p> element to display the phrase "Hello World!" on the display screen.

The last line closes the second card.

The Deck Footer

Our deck ends with a closing deck element:

```
</wml>
```

This will always be the last piece of code on any WML deck that you create.

A SERVICES SITE EXAMPLE

The preceding "Hello World" site may look simple, but it contains the very basics of WML, and you can go ahead and create a very nice, simple, static WAP-enabled site by applying only the preceding tags. The rest of this chapter will explain what you can do to make a really nice site.

Knowing how to markup text with HTML is an advantage, and you might also know about the Internet client/server model, and how a client (browser) makes requests to the server, which then responds by sending out the appropriate Web page. You may even have worked with dynamic server technologies, such as ASP or CGI. However, this knowledge does have some disadvantages. Translating the HTML concepts across to WAP and WML is easy enough, but the practicalities can be a little trickier because of the basic restrictions that a WAP device must contend with as described at some length in earlier chapters: which are small screen size, small monochrome images for graphics, and slow data-transfer speed.

Because WML is compiled into a binary format, it is totally unforgiving of coding errors such as incorrectly nested tags, using uppercase characters for tags, and so on. And by unforgiving, I do not mean that it gives an error message. I mean that the user sees a blank screen instead of the content you intended.

Using Multiple Decks

We have already introduced cards and decks, so the question is now "How do I get all of my data into a single deck? That's going to be a lot of cards, isn't it?" The simple answer is that we can have any number of decks, each with its own collection of cards. This is where we get to the design of the application, and where we split the cards into a sensible navigable structure.

Each deck can also contain a "template" structure that tells the micro-browser what to do if the user presses a key (for example the "Back" key), so we should design this to handle the default actions that we want to take *in that particular deck*.

Let's take a look at an example site for a company that designs Web pages and sites. The site contains a welcome page, help page, and contact page in the first file (deck), and then three product/service pages in a second file (deck).

The first file, "main.wml", contains the following:

- ▼ Template   Help, Back, and Home buttons
- ■ Card 1   Welcome page
- ■ Card 2   Help page
- ▲ Card 3   Contact page

The second file, "products.wml", contains the following:

- ▼ Template   Help, Back, and Home buttons
- ■ Card 1   Standard Site Design service page
- ■ Card 2   Interactive Database-Driven Site Design service page
- ▲ Card 3   E-Commerce Site Design service page

The whole site fits in two files, or decks, but it contains six screen pages, or cards.

In a way, the deck concept is somewhat similar to the way you can use the # label anchor in HTML to break up a large page into different navigable sections. The only difference is that, typically, only one # section, or card, is viewed at a time by the WAP device.

Building the Services Site

To build the Web page design site we've just outlined, we must create the first .wml file. The skeleton of the "main.wml" file looks like this:

```
<?xml version="1.0"?>
<!DOCTYPE wml PUBLIC "-//WAPFORUM//DTD WML 1.1//EN"
"http://www.wapforum.org/DTD/wml_1.1.xml">

<wml>

<template>
</template>

<card title="Welcome" id="main">
</card>

<card title="Help" id="help">
```

```
</card>

<card title="Contact us" id="contact">
</card>

</wml>
```

Inside the First Deck

Inside the `<wml>` tags are the `<template>` and `<card>` sections. These are basically the only things a deck can contain, and anything outside of these sections will cause an error.

The order of the cards in the deck doesn't matter, as long as the first one that you want the user to see is at the top. (Naturally, as a good coder, you would lay out the cards in a logical and sensible order simply for the sake of good code readability, so that you can easily make those urgent modifications a year from now.) Each card has been given its own identity with the `id` attribute, so that once we start linking them, they can all be identified unambiguously.

The template section describes the layout of the buttons that will be common across all of the cards. Commonly a Back button is assigned to every card, so that the user never gets "dead-ended" on a card with no way forward or back. (We'll consider this in more detail later.)

Finally, remember that all tags and attribute names should be in lowercase only. Unlike HTML, this is mandatory, and will generate an error when the cards are compiled if you get it wrong.

What's in a Card

What can actually go in a card? Simply put, a card should be a discrete piece of information that fits nicely on the screen of a mobile device and that can be viewed without too much, if any, scrolling up and down.

A card, for example, could contain a small text article (such as a weather report or a daily horoscope), a picture (black and white only), a menu, or a question from a form (one question per card).

On a text card, *all* text is formatted and displayed in paragraph tags. In HTML, you have probably only used the paragraph (`<p>`) tag occasionally around text, and mainly to add a gap between sections of text on a page. If you type HTML text directly on the page, the browser will display it as it thinks best.

This is very nice and easy, but it won't work with WML. The WML rules are simple:

▼ All text has to be placed inside paragraph tags.

■ All paragraphs have to start with a `<p>` tag and end with a `</p>` tag.

▲ Paragraphs cannot be nested within each other.

Let's start with the Help card of our main.wml code skeleton. A simple card might look like this:

```
<card title="Help" id="help">
<p>
WebDesigns provides attractive web sites for small to medium sized companies.
</p>
<p>
This site contains details of our prices: click the prices link.
</p>
</card>
```

Here we have two paragraphs, both of which have a sentence of text in them. This would be displayed on a micro-browser as shown in Figure 5-3.

As you can see, even single sentences can take up all the screen space, so you should write your text with that in mind.

Graphics

The first thing that most people will want to do with a WAP site is to put some kind of splash screen or graphic up for the user to see. Despite the limited bandwidth available for WAP, graphics are not going to go away, so let's see how to do it for WAP.

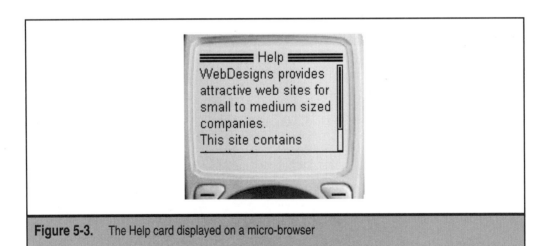

Figure 5-3. The Help card displayed on a micro-browser

CAUTION: These splash screens are usually justified as "branding," or "corporate image," or something like that. Unless they are created with a small, rapidly downloadable graphic that actually conveys a message, I consider it a waste of the user's valuable time or a gross misuse of bandwidth, but that's just my opinion. If users refuse to use your application because it takes too long to load, however, then it becomes your problem.

An image created for use with WAP is in a special format called WBMP (WAP bitmap), which is a bitmap image for WAP devices. These images are not the latest and greatest in multimedia formats, but they can still be useful for a corporate logo, or as we saw in Chapter 1, for a stylized map.

To design the graphic, you will need a graphics editor that can output WBMP images. Paint Shop Pro (available from http://www.jasc.com) can create and save images in this format, and there will no doubt be others by the time you read this. You can also try Teraflops' online WBMP converter, which is at http://www.teraflops.com/wbmp/.

A WBMP graphic is only a monochrome image, but with a little bit of effort, (in the case of the following illustration, a *very* little bit of effort) something resembling a logo can be achieved.

This particular image was created in Paint Shop Pro in monochrome and saved as a WBMP image. The total file size of the completed image shown here is only 354 bytes.

The <img> Tag

Once you have created the graphic and saved it (in this case as wd.wbmp), you can use WML's `<img>` tag to support and display images, much as you do in HTML.

There are two differences between the use of the HTML and WML `<img>` tags. Firstly, the `alt` attribute of the `<img>` tag is mandatory in WML. This tells mobile devices that can't currently support images exactly what text to put in their place. Secondly, the `<img>` tag is one of a few in WML that do not have a closing tag. It therefore needs to have its own closing mark, a slash (/), actually *inside* the tag, like so:

```
<img src='wd.wbmp' alt='WebDesigns Logo'/>
```

If we save our WBMP file in the same directory as the WML page, the card that displays the graphic will look like this:

```
<card title="Welcome" id="main">
<p align='center'>
<img src='wd.wbmp' alt='WebDesigns Logo' />
</p>
</card>
```

Notice how I used the "align" attribute of the `<p>` tag to get the logo in the center of the screen. The resulting output is shown here:

We have now written two very simple cards to create a deck, and we have created a graphic "work of art." As you can see, it is not rocket science that we are using here.

The Services Site with Graphics

Now let's start building on what we have covered so far. Let's create both decks that we described earlier, and see how we can navigate between them. By the time we finish, we will have a working WAP site that you can post on the Web and use to send my company, WebDesigns, lots of extra work.

Here is the code for the main.wml deck:

```
<?xml version="1.0"?>
<!DOCTYPE wml PUBLIC "-//WAPFORUM//DTD WML 1.1//EN"
 "http://www.wapforum.org/DTD/wml_1.1.xml">

<wml>

  <template>
  </template>

  <card title="Welcome" id="main">
    <p align='center'>
      <img src='wd.wbmp' alt='WebDesigns Logo'/>
    </p>
  </card>
```

```
  <card title="Help" id="help">
    <p>
      WebDesigns Ltd builds unique web sites for all kinds of companies. We know your company is unique too.
    </p>
    <p>
      This site contains details of our creative services: click this link.
    </p>
    <p>
      To view more services, click the 'Back' button to return to the main menu.
    </p>
  </card>

  <card title="Contact us" id="contact">
    <p>
      You can reach us anytime on 777 777 7777
    </p>
  </card>

</wml>
```

Here is the products.wml deck:

```
<?xml version="1.0"?>
<!DOCTYPE wml PUBLIC "-//WAPFORUM//DTD WML 1.1//EN"
"http://www.wapforum.org/DTD/wml_1.1.xml">

<wml>

  <template>
  </template>

  <card title="Our Services" id="services">
    <p align='left'>
      Web Sites
      <br/>
      Data Driven Sites
      <br/>
      E-Commerce Sites
    </p>
  </card>

  <card title="Web Sites" id="websites">
    <p align='center'>
      Standard Web Sites
    </p>
    <p align='center'>
      Get a Quote!
    </p>
  </card>
```

```
<card title="Data Driven Sites" id="datadriven">
  <p align='center'>
    Interactive Database-Driven Sites
  </p>
  <p align='center'>
    Get a Quote!
  </p>
</card>

<card title="E-Commerce Sites" id="ecommerce">
  <p align='center'>
    Full Scale E-Commerce Sites
  </p>
  <p align='center'>
    Get a Quote!
  </p>
</card>

</wml>
```

Now that we have built the basic structures, we need to flesh them out and add the links that will "drive the site." Without the links added, all you will see if you open either of these decks on a WAP device are the first cards.

Creating Links

If you are already a Web developer, you'll be pleased to know that links are provided by the familiar `<a>` tag. The `<a>` tag has an attribute, `href`, that points to the destination, like so:

```
<a href='products.wml'> Link Description to be displayed here </a>
```

For the `<a>` tag, you include the description of the link and you must close the tag as well, as shown.

This code would provide a link into the products deck, assuming it was in the same directory as the card that supplied the link. The `href` attribute is exactly the same as in HTML links; it can be absolute or relative. The above example is using a relative link, "relative" meaning relative to the location of the file that is doing the calling. The above example assumes that the file "products.wml" is located in the same directory.

The following example calls the same file, but uses an "absolute" reference, in that it precisely locates the file on the whole network, and it cannot be confused with any other file in the world.

```
<a href='http://www.webdesigns.ltd.uk/wml/products.wml'>
```

These links will take the micro-browser to the top of each deck. If you want to link to a specific card, you should use the hash (#) fragment of the URL as shown here. (This is identical to the label anchor tag in HTML):

```
<a href='http://www.webdesigns.ltd.uk/wml/products.wml#ecommerce'>
```

This would take you to the details of the E-Commerce site-creation service.

The hash fragment can be used by itself to link to another card within the same deck, like so:

```
<a href='#ecommerce'>
```

This hash fragment links to another card in the same deck.

The entire deck has already been downloaded to the user's WAP device when the deck was first linked to. So, calling another card in the same deck will produce a virtually instantaneous response, and the user will be most grateful for your thoughtfulness as a developer.

The WML Site with Links

Our site requires links from the "Welcome" page to the "Help" page, and from the "Welcome" page to the "Contact" and "Product" pages. The product list then needs to link to each of the products themselves. (We'll leave the "Get a Quote" links for later.)

The full WML for the site, so far, follows. First, here is the "main.wml" deck:

```
<?xml version="1.0"?>
<!DOCTYPE wml PUBLIC "-//WAPFORUM//DTD WML 1.1//EN"
"http://www.wapforum.org/DTD/wml_1.1.xml">

<wml>

  <template>
  </template>

  <card title="Welcome" id="main">
    <p align='center'>
      <img src='wd.wbmp' alt='WebDesigns Logo'/>
      <br/>
      <a href='#help'>
        Continue...
      </a>
    </p>
  </card>

  <card title="Help" id="help">
    <p>
      WebDesigns Ltd builds unique web sites for all kinds of companies. We know your company is unique too.
    </p>
    <p>
      This site contains details of our creative services: click this <a href='products.wml'>link</a>
    </p>
    <p>
      To view more services, click the 'Back' button to return to the main menu.
    </p>
```

```
    </card>

    <card title="Contact us" id="contact">
      <p>
        You can reach us anytime on 777 777 7777
      </p>
    </card>

</wml>
```

And here is the "products.wml" deck:

```
<?xml version="1.0"?>
<!DOCTYPE wml PUBLIC "-//WAPFORUM//DTD WML 1.1//EN"
"http://www.wapforum.org/DTD/wml_1.1.xml">

<wml>

  <template>
  </template>

  <card title="Our Services" id="services">
    <p align='left'>
      <a href='#websites'>
        Web Sites
      </a>
      <br/>
      <a href='#datadriven>
        Data Driven Sites
      </a>
      <br/>
      <a href='#ecommerce'>
        E-Commerce Sites
      </a>
    </p>
  </card>

  <card title="Web Sites" id="websites">
    <p align='center'>
      Standard Web Sites
    </p>
    <p align='center'>
      Get a Quote!
    </p>
  </card>

  <card title="Data Driven Sites" id="datadriven">
    <p align='center'>
      Interactive Database-Driven Sites
    </p>
```

```
      <p align='center'>
        Get a Quote!
      </p>
    </card>

    <card title="E-Commerce Sites" id="ecommerce">
      <p align='center'>
        Full scale E-Commerce Sites
      </p>
      <p align='center'>
        Get a Quote!
      </p>
    </card>

</wml>
```

The first card of the first deck now has a Continue link at the bottom, and the service index (shown in the following illustration) has links to each of the service cards.

Templates

The final thing we need to cover at this point is the `<template>` tag, which allows you to set default characteristics for buttons or events that are common to cards across the whole of a deck.

As mentioned earlier, having a Back button that appears on all the pages of a deck is a very common requirement. The template code for implementing a Back button would be as follows:

```
<do type='prev' label='Black'>
  <prev/>
</do>
```

The `<prev>` tag is an action, and it tells the phone to go back to a previous card in its history. The `<do>` tag suggests to the phone that the label "Back" should be placed above the soft key that is normally used as the default for the "Prev" or "Back" action. I say "suggests," because no two mobile devices are the same or even implement the same code in the same way. As already mentioned at some length, there is no way to guarantee the appearance of any deck on any device without extensive testing on specific devices, and even using different versions of micro-browsers on the same device.

To place the preceding Back button in every card of the deck, just place the `<do>` tag in the `<template>` tag at the top of the deck, like so:

```
<template>
  <do type='prev' label='Back'>
    <prev/>
  </do>
</template>
```

Now the Nokia micro-browsers will display something like what is shown in the following illustration for the preceding WML code.

A Phone.com micro-browser, however, looks something like the following:

Every product card has its own Back button, and the user can't get stuck or "dead-ended." You now have a fully functional WAP site, and you can use this as a starting point to build on.

The next thing we need to look at is how to get data back from the user. Let's take a look at that in the next chapter.

CHAPTER 6

Interactivity: Forms and User Input

Collecting data from the user is something that we have to do in any application, regardless of the content. Giving the user the option to select a choice from a menu of items and then collecting the data on which option was selected is a basic part of this. The entire user interface and the interaction of the application with the user are part of collecting data from the user and then performing an action within the application based on that user input.

First of all, we are going to look at the different options for allowing the user to make straight choices between items. These are usually in the form of menus and submenus, allowing users to drill down to the exact data that they want.

THE OPTIONS MENU (SELECT)

The first of the methods that we are going to look at for doing simple selections between different options is, appropriately enough, the **options** statement. Here is an example showing a simple selection of an animal that the user would like to have as a pet:

```
<?xml version="1.0"?>
<!DOCTYPE wml PUBLIC "-//WAPFORUM//DTD WML 1.1//EN"
"http://www.wapforum.org/DTD/wml_1.1.xml">
<wml>
<card id="FirstCard" title="First Card">
<p align="center">
Pick a Pet
<br/><br/>
<select name="x" title="Pets">
   <option value="c">cat</option>
   <option value="d">dog</option>
   <option value="g">guinea pig</option>
   <option value="h">hamster</option>
   <option value="s">snake</option>
</select>
</p>
</card>
</wml>
```

All we are doing here is displaying the line "Pick a Pet," and then allowing the user to select one of several animals. (If this were a quiz, the correct answer would of course be cat, as cats are the undisputed masters of the universe.) In the following sections, you'll see what the resultant screens look like on the Nokia and Phone.com micro-browsers.

Selection on the Nokia

When first loading the deck on the Nokia, you see the default selected option—in this case, "cat":

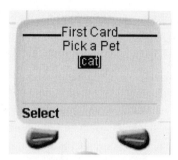

Pressing the Select soft key takes you into the coded **select** statement, and you will see the options with radio buttons next to them. If you scroll down to "guinea pig" and press the Select soft key, you will get the following screen. Note that you have to press the Select key to get the radio button to appear in the correct option.

If you now press the OK soft key, you are taken back to the first screen, but with "guinea pig" selected, like so:

Selection on Phone.com

The code for our pet selection program produces the following screen on the Phone.com micro-browser:

With the Phone.com micro-browser, you may select the animal by scrolling down the list and pressing the OK soft key, or you can press the number next to the item. To select "guinea pig," for example, you press the 3. That is all you need to do with this browser.

The above code doesn't actually do anything very useful at the moment but is simply intended to demonstrate the use of the **select** statement. We will build on these statements to produce something workable as we go.

Option Groups

Option groups (the **optgroup** element) allow you to group multiple **option** statements, just as they are used in the **select** statement, within a card. You can nest these, thereby allowing a lot of scope for items with multiple sublists.

At the time of writing, the **optgroup** element is only usable on the Nokia micro-browser, *not* on the Phone.com micro-browser. If you have a specific target phone that has the Nokia micro-browser, you are fine. If not, you may as well ignore this element for the time being.

How the **optgroup** element works is like this. First of all, make a list of all of the items that you want to display, in their appropriate subgroups, like so:

Building materials

- ▼ Bricks
    - ■ Standard
    - ■ Blocks
    - ■ Custom
- ▼ Concrete
    - ■ Grade 1

- Grade 2
- Grade 3

▼ Glazing
- Standard
- Double
- Triple

Now you simply add the **select**, **optgroup**, and **option** statements that apply to each item, like so:

```
<select title="Select to Order">
<optgroup title="Bricks">
   <option>Standard</option>
   <option>Blocks</option>
   <option>Custom</option>
</optgroup>
<optgroup title="Concrete">
   <option>Grade 1</option>
   <option>Grade 2</option>
   <option>Grade 3</option>
</optgroup>
<optgroup title="Glazing">
   <option>Single</option>
   <option>Double</option>
   <option>Triple</option>
</optgroup>
</select>
```

Because all of the items are displayed in the order that you specify, this can be a powerful tool for customizing sales order items for sales personnel who are on the road a lot.

If you create a deck with the above code in it and run it on a Nokia micro-browser, you will see the following:

If you select Glazing, and then pick Double, you will see this:

A major benefit of the **optgroup** element is that although we had nine separate items in three groups, we could fit them all onto the screen in groups of three. This means that in this particular case, users don't have to scroll off the end of the screen, and they are led in a very intuitive way to the data that they need.

To make this even more useful, you can use the **onpick** attribute of the **option** element to jump directly to another URL or card. In the above case, we could add more cards to give current prices of each item, or a fuller description, or whatever we like.

I will show you all of the code for the deck to give you the idea. The items in bold are those that I have added to the preceding code.

```
<?xml version="1.0"?>
<!DOCTYPE wml PUBLIC "-//WAPFORUM//DTD WML 1.1//EN"
"http://www.wapforum.org/DTD/wml_1.1.xml">
<wml>
    <card id="Products" title="Building Materials">
    <p>
    <select title="Select to Order">
    <optgroup title="Bricks">
        <option>Standard</option>
        <option>Blocks</option>
        <option>Custom</option>
    </optgroup>
    <optgroup title="Concrete">
        <option>Grade 1</option>
        <option>Grade 2</option>
        <option>Grade 3</option>
    </optgroup>
```

```
        <optgroup title="Glazing">
           <option onpick="#glaz1">Single</option>
           <option onpick="#glaz2">Double</option>
           <option onpick="#glaz3">Triple</option>
        </optgroup>
        </select>
        </p>
        </card>
        <card id="glaz1" title="Single Glazing">
        <p>
           Heat Loss Coeff= 0.75<br/>
           <b>10.00 per sq.mtr.</b>
        </p>
        </card>
        <card id="glaz2" title="Double Glazing">
        <p>
           Heat Loss Coeff= 0.5<br/>
           <b>20.00 per sq.mtr.</b>
        </p>
        </card>
        <card id="glaz3" title="Triple Glazing">
        <p>
           Heat Loss Coeff= 0.5<br/>
           <b>40.00 per sq.mtr.</b>
        </p>
        </card>
</wml>
```

If you enter the above code, then select Glazing and then Double from the menu options, you will see the following screen:

You can play with the code to see how it works. Later on, when we get to the point of pulling in data from databases, you will see how easy it is to keep an application like this up-to-date with current data, prices, and so on.

NOTE: Incidentally, I have absolutely no idea whether the previous figures mean anything at all, as I just made them up for the example. I remember the term "Heat Loss Coefficient" from my school days, and it's the first time I have ever used it in real life.

If you do run this code on a Phone.com micro-browser, it will run, but it will only show the **option** statements, not the **optgroup** statements. The **onpick** attributes still work, but the whole grouping concept is lost. So when you first load the deck, it looks like the following:

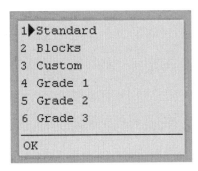

TEMPLATES REVISITED

We covered the concept of the **template** element in the previous chapter. As you will know if you tried the code earlier, you got stuck, or dead-ended, in every screen that you ended up in. If you remembered the **template** element and tried adding it, as shown in the previous chapter, this would not have happened to you. If you are like me, and need several reviews of new information before it penetrates, don't worry, the review starts now.

The **template** element allows you to set default characteristics for buttons or events that are common to cards across an entire deck. The **template** code to implement a Back button would be as follows:

```
<template>
  <do type='prev' label='Back'>
    <prev/>
  </do>
</template>
```

The **prev** tag is an action, and it tells the phone to go back to a previous card in its history. The **do** tag suggests to the phone that this should go on the "prev," or Back, button. Now every product card in that deck has its own Back button, and the user can't hit a dead end. And remember, you have to add the **template** tag to every deck you create, not just the first one.

Try adding the above **template** code to the code we have already created, and see what happens when you test it. When you get to the details of the type of glazing you want, you now have a Back button that will take you back to the **select** option.

But the **template** element is more versatile than just preventing the user from getting stuck because you forgot to add a `<do type='prev' label= 'Back'> <prev/></do>` statement to a card. Let's say you have three or four menu options that you want the user always to be able to select from, with just two clicks. This means that you have to put the menu onto one of the soft keys of the mobile device, and you have to add the menu code to every *card*. But if you put the same code into a **template** element, and add this code once to the top of each *deck*, then the same effect can be accomplished without any of the extra effort or code maintenance that would be needed otherwise.

Just imagine you have written an application consisting of numerous decks and cards, including all of the menu code on each card, and then the boss says something like, "I don't like the word 'help' on the menu, so make it read 'assistance'. Ouch. Ouch. Ouch. The "Ouch. Ouch. Ouch" is in your imagination as you picture yourself hitting your boss over the head with your keyboard. So spare your boss a trip to the hospital— use a template.

Let's see how this would work. We want to add the menu options Help, Home, and Contact Us to every card, so that no matter where users are in the application they can always access these options with the same text. We can use the **template** element along with the **do** element to write something like this:

```
<template>
    <do type="accept" label="Home">
    <go href="menu.wml#home"/>
    </do>
    <do type="accept" label="Help">
    <go href="menu.wml#help"/>
    </do>
    <do type="accept" label="Contact Us">
    <go href="menu.wml#contact"/>
    </do>
</template>
```

(There are several different **type** actions that we can use. We have already used **prev**, the one used here is the **accept**, and we will cover the rest in the next section.)

WAP: A Beginner's Guide

> **NOTE:** The Nokia SDK version 2.0 doesn't like this **template** statement. It says that there are one or more **do** elements with the same name. The Phone.com SDK accepts it without errors, and it certainly works, but this is all part of the "fun" of being out on the bleeding edge of technical development.

The Phone.com browser takes the preceding **template** statement and produces the following screens. The first option, Home, is placed on one soft key, and the title Menu is placed over the other soft key. Pressing the Menu key displays the remaining two options, Help and Contact Us.

The Do Element

As mentioned in the previous section, there are a few more **type** attributes to the **do** element that are worth knowing about, aside from the **accept** attribute that normally equates to the OK, or Yes, button. Table 6-1 lists the attributes and a brief description.

Type	Description
prev	Previous, or Back button.
help	The button pressed by the user to get help.
reset	Reset the device or application.
options	Request for more operations.
delete	Delete or remove an item.
unknown	You can map this to any key on the device.

Table 6-1.   The Do Element "Type" Values

Where and how these items appear when used depends entirely on the type of device that is displaying the card, but let's see some examples.

The Help Type

This is the button or soft key pressed to request help from the application.

```
<wml>
    <template>
      <do type="help"><go href="#help"/></do>
      <do type="prev"><prev/></do>
    </template>
<card id="MainCard" title="Main Card">
<p>
This is some card one text
</p>
</card>
<card id="help" title="Help">
  <p>This is some application help!</p>
</card>
</wml>
```

The code produces the following screens. On the left is the main screen, and when the Help soft key is pressed, the user sees the screen on the right.

You will notice in the screen on the right that the Help soft key is still showing. This is a good place to introduce the **noop** statement. The actions of the **do** element can be overridden at the card level. In this particular case, we do not want to display the Help action, so we can override this at the card level by inserting the following statement in the Help card.

```
<card id="help" title="Help">
  <do type="help"><noop/></do>
  <p>This is some application help!</p>
</card>
```

When this is done, the Help screen will look more professional, quite aside from being more logical:

The Reset Type

This type does exactly what it says. You should always allow your code to close down "nicely," or in a completely controlled fashion; otherwise it can produce random side effects. I'm not sure when I might actually use this type, as it dumps the whole deck without preamble. I would probably label this option "For Emergency Use Only," and then make sure that such an emergency never occurs.

The Options Type

You can use the **options** type to create more menus of items, or to carry out a specific action that you want to label separately as an option. It is always good to have another shortcut!

In fact, you can create any kind of type that you like and assign any kind of action to it. You can create a **do** element with a user's name on it and a specific action to take that user to a frequently referenced page with, say, phone numbers on it. For example:

```
<wml>
    <template>
    <do type="prev"><prev/></do>
    <do type="Secret"><go href="#card3"/></do>
    </template>
      <card id="card1" title="First Card">
      <p align="center">
        <big><b>First Card</b></big>
      </p>
    </card>
      <card id="card2" title="Dummy Secret">
      <do type="Secret"><go href="#card2"/></do>
      <do type="prev"><go href="#card1"/></do>
      <p>John: 111-1111</p>
      <p>Anne: 222-2222</p>
```

```
    <p>etc...</p>
  </card>
  <card id="card3" title="Secret">
    <do type="Secret"><go href="#card2"/></do>
    <p>Spouse: 555-5555</p>
    <p>Lover: 666-6666</p>
    <p>Florist: 777-7777</p>
  </card>
</wml>
```

This produces a set of screens beginning with this one:

Pressing the Secret option produces the screen shown next on the left, and pressing Secret again (if somebody tries to look over your shoulder!) will show the screen on the right.

In this case, on the right-hand screen I have overridden both of the **template** elements, and so pressing Secret again will only show the same screen; pressing Back will take you back to the first screen only, not the "real" secret screen.

There are other ways to control the flow of the screens that we will come to shortly, but this little joke scenario gives you the idea of the kind of things that can be done with some very simple tags. The point is that you can use the **do** element as another way of creating menus of options for the user to select from.

EVENTS

"Event" in ordinary language can be defined as "something happens." As far as we are concerned, in programming "event" is identical in meaning, but with one major difference. When "something happens" in a computer system, the system itself has to (1) detect that something has happened and (2) know what to do about it.

In real life, if you put your hand on a hot stove, the "operating system" takes over and (1) snatches your hand away and (2) screams. In a computer system, the program is chugging away merrily when the user clicks on a mouse button. The operating system has to detect that a mouse button has been clicked (the event) and then do something about it, which is known as the *event handler*. In normal usage this all looks pretty smooth, but if the programmer doesn't set up the event handler for a specific event, nothing will happen when that event occurs.

A graphical user interface like Windows is chock full of these events, which are happening all the time—mouse moves, mouse pointer location (with coordinates), keyboard keypress, mouse button press, mouse move while button pressed, and so on and on. What is physically happening inside the system is that the operating system is detecting these events, assigning them relative priorities, and then queuing them for handling by their respective event handlers in the correct sequence.

Fortunately, we are not concerned with a great many events or event handlers with WAP at this point because the user interface does not include such dynamic things as a mouse—yet! But a major advantage of specifying events and event handlers is that we can make an application behave dynamically. That is, we can guide users to where we want them to go depending on what has been done already, rather than depending on users going back up a menu structure and then down again to the option that is needed. Say, for example, we want users to place an order for something that they have already picked in a separate module of an e-commerce WAP application. The events that we can specify for handling when they occur are

- onenterbackward
- onenterforward
- onpick
- ontimer

Remember that these must be lowercase! Let's look at the four events one at a time.

Onenterbackward

The **onenterbackward** event occurs when the user hits a card by normal backward navigational means. That is, he presses the Back key on a later card and arrives back at this card in the history stack. In many instances this would probably be just fine, but there are times when you want to redirect the user to another card or deck.

Chapter 6: Interactivity: Forms and User Input

As a simple example, say the user has entered his password in clear text on one screen; so you wouldn't want somebody else to be able to pick up his phone later and press the Back key a few times and learn the password. Without the **onenterbackward** event being defined, you might have a simple deck that looks like this:

```
<wml>
    <template>
        <do type="prev"><prev/></do>
    </template>
<card id="Entry" title="Password Screen">
<p>
  <input name="pass"/>
  <a href="#welcome">Login</a>
</p>
</card>
<card id="welcome" title="Welcome">
  <p>Welcome to the main screen.
  <br/>
  <a href="#help">Help</a>
</p>
</card>
<card id="help" title="Help">
  <p>Press any key for help on that key...</p>
</card>
</wml>
```

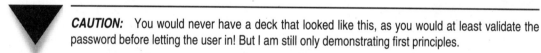

CAUTION: You would never have a deck that looked like this, as you would at least validate the password before letting the user in! But I am still only demonstrating first principles.

Now let's take a look at the screens you see when you run this deck. Here is the first screen:

Clicking on the Edit soft key that automatically appears on this screen in the Nokia browser gives you the screen shown on the left in the following. Entering my name in this edit screen and then clicking the OK soft key brings up the next screen, shown on the right.

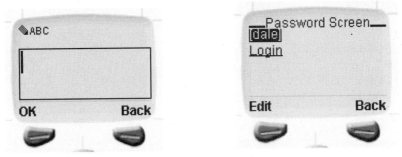

Selecting the Login option causes the soft key to automatically change to Link, as shown next on the left. And clicking on the Link soft key will take you to the Welcome page (shown in the right-hand screen), where there is another link to a Help screen.

The problem now arises that if you press the Back soft key, you will be taken back to the login screen with the password showing. But if we now add the event handler **onenterbackward** to the login card, we can redirect the user to another location—in this case back to the Welcome screen, making the Welcome screen the "new" first card in the deck for this session.

The revised code for the first card would now look like this:

```
<card id="Entry" title="Password Screen">
  <onevent type="onenterbackward">
    <go href="#welcome"/>
  </onevent>
<p>
  <input name="pass"/>
  <a href="#welcome">Login</a>
</p>
</card>
```

Now, wherever you go in the application, pressing Back as many times as you like will only ever take you back to the Welcome screen.

Finally, now that you have seen this in "long" form as

```
<card id="Entry" title="Password Screen">
  <onevent type="onenterbackward">
    <go href="#welcome"/>
  </onevent>
```

you should note that this **onevent** statement could also be written in a shorthand form as

```
<card id="Entry" title="Password Screen" onenterbackward="#welcome">
```

This is not only a large saving in bandwidth, but also easier to type and more logical in construction, as the events all apply to a card in any case.

This shortcut applies to all of the event handlers, and they can also be used in the **card** element in any combination.

Onenterforward

The **onenterforward** event occurs when the user hits a card by normal forward navigational means. This is not as commonly used as **onenterbackward**, but there are times when it can be useful. For example, if you want to initialize one or more variables when you first enter the application, but not when you return to the same card by going back to it. You could write this as:

```
<card>
   <onevent type="onenterforward">
      <refresh>
         <setvar name="country" value="United Kingdom">
      </refresh>
   </onevent>
   <p>
   Country:
   <br/>$(country)
   </p>
</card>
```

You can also use the **onenterforward** and **onenterbackward** events together to control the flow of the application. The code would look something like this:

```
<card id="main" onenterforward="#main2" onenterbackward="#login">
</card>
<card id="main2">
   ...
```

```
</card>
<card id="login">
...
</card>
```

NOTE: You can place the event types directly into the card element as shown in the previous code.

Onpick

The **onpick** attribute is a great shortcut if you are using a **select** menu. Instead of writing a lot of code that allows the user to go to another card if an option is selected, you can simply place the destination into the **onpick** attribute. Here is a code fragment without the **onpick** attribute:

```
<card id="beds">
<p>
    Select the number of bedrooms
    <select title="Houses">
    <option>3 Bedroom</option>
    <option>4 Bedroom</option>
    </select>
<a href="#view">View</a>
</p>
</card>
<card id="view" title="View House">
<p>
    Are you sure you want to view this house?
    ...
</p>
</card>
```

and here is the equivalent code, only this time using the **onpick** attribute:

```
<card id="beds">
<p>
    Select the number of bedrooms
    <select title="Houses">
    <option onpick="#bed3">3 Bedroom</option>
    <option onpick="#bed4">4 Bedroom</option>
    </select>
</p>
```

```
</card>
<card id="bed3" title="3 Bed House">
<p>
   Nice house. Are you sure you want to view?
   ...
</p>
</card>
<card id="bed4" title="4 Bed House">
<p>
   Big house!  Are you sure you want to view?
   ...
</p>
</card>
```

This is a tremendous time saver, and one that you will find yourself using over and over.

Ontimer

The use of timers on a site may seem a little strange at first. After all, the data is the data, right? Well, this will depend on the application that you want to write and the kinds of features that you may wish to add. While the most common current use of the **timer** event is to display a splash screen, it can also be used to refresh a page of information that is likely to change at irregular intervals, such as stock prices, without the user having to think about it. The usage is very simple—you just need to specify an action to be performed and a time value in tenths of a second.

NOTE: Although splash screens are much favored among web site designers and web site owners, they are rated as one of the ten things most likely to irritate regular users of online services of any kind.

The following block of code will display a splash screen for five seconds, and then move on to the welcome screen in the deck if the user has not pressed the Enter link in that time. (This is an example only—please don't "do this at home" unless you enjoy irritating users.)

```
<card id="splash">
  <onevent type="ontimer">
      <go href="#welcome"/>
  </onevent>
  <timer value="50"/>
<p>
   <a href="#welcome">Enter</a>
</p>
</card>
```

```
<card id="welcome" title="Welcome">
  <p>Welcome to the main screen.
  <br/>
  </p>
</card>
```

The shortcut for the first card could be written as

```
<card id="splash" ontimer="#welcome">
  <timer value="30"/>
<p>
    <a href="#welcome">Enter</a>
</p>
</card>
```

VARIABLES

In most other computer languages, you can create variables that hold all sorts of data, such as character strings, numbers, and so on. You normally do this by declaring a variable name and then stating what type of data the variable is going to hold. For example, the command

`dim x as char 20`

declares a variable called "x" that is going to hold a character string 20 characters long.

You are also usually able to declare a variable in a section of the program so that it cannot be seen by any other sections of the program.

This is all extremely useful. You can ask the user for her name and store this in a variable so you can then call the user by name at some future point in the program. For example:

`x = "Jane Smith"`

And then in another place in the program, you can write something like `Hello $x`, and the program will print `Hello Jane Smith`.

NOTE: The $ sign is one example of how the program knows that you are referring to a variable and not just to the letter *x*. The syntax of exactly how this is done will vary from one computer language to the next.

The ability to have a variable that is not visible in another part of the program means that even in medium-sized applications you can use a variable called "x" in several places if you want to, without one version interfering with another. But even while taking care that this does not happen, it sometimes does occur and can create some very esoteric bugs, particularly when you're dealing with numbers. (Not that you would ever have several versions of a variable called "x", of course!)

In WML, variables are both good news and bad news. The good news is that they're very simple to use. The only type that they can be is **character**. They're easy to declare, and they're easy to assign values to. The bad news is that any WML variable is "global" to the entire WML environment. What this means is that if you use the variable "x" to store a piece of data in one deck, and then use the variable "x" in another deck to store another piece of data; when you go back to the first deck, a variable "x" will now contain what was entered in the second deck. The really bad news is that this will apply even if the "second deck" was written by somebody else, or belongs to some other application altogether.

What this means to you, as the developer, is that you have to make very sure that you initialize all variables correctly when they are needed, or in the deck that they are needed, and that you keep your naming conventions consistent to avoid confusion.

There is an attribute of the **card** element called **newcontext** that will not only destroy all variables, but also the entire browser history. The only time you might want to use this would be when you want to specifically ensure that the entire "slate is wiped clean" as far as the micro-browser is concerned. But this, once more, is a last-ditch scenario, or something that you would put on the very first card of your application if you had to. Even using it on the first card of your application is not a particularly good idea because you don't know how somebody has arrived at your application. If you were a user who had followed a data trail to your application, and then you pressed the Back button and couldn't get back where you were before, would you be a happy user? It is little things like this that make users wary of your application, or unwilling to return in the future. Put yourself in the user's shoes and see how willing you would be to have something like this happen to you.

Using Variables

The only thing you really need to know about using variables in the beginning is that variable names can contain only letters, numbers, or an underscore character, and they can only start with a letter or an underscore character. So, for example, the following three variables are all fine:

```
deck_one_cat,
_deck1_cat
deck53_card4_cat
```

But these two will fail:

```
1deck_cat
deck1*cat
```

To create, or set, a variable, you use the **setvar** tag. The **setvar** tag only has two attributes—**name** and **value**. These are, simply, the name of the variable, and the value that you wish to assign to it. For example, the following creates a variable called "test" with a value of "hello."

```
<setvar name = "test" value = "hello">
```

In order to use this variable, all you have to do is prefix the variable name with a $. When the micro-browser finds a $, the variable name is simply replaced with the contents of the variable. The only twist to this is when you want to use the $ as an actual currency symbol. In this event, all you have to do is to type two $ $, and this will be interpreted as a single $.

The **setvar** tag can only be used in a few places. It can be used inside the **go** and **prev** tags, which we have already covered, and in a new one, **refresh**. The **refresh** tag causes the variable assignment within it to be redone every time the card is displayed, and it is most commonly used for setting variables.

The **refresh** tag itself can only be used within an **onevent** or a **do** element. Here is a very simple example:

```
<wml>
    <card id="main" title="Show Variable">
      <onevent type="onenterforward">
        <refresh>
           <setvar name="var_name" value="Mr King"/>
        </refresh>
      </onevent>
      <p>
         Welcome to my application, $(var_name).
      </p>
</card>
</wml>
```

This code produces the following screen:

TIP: You can use punctuation around the variable name to make the resultant output look more natural, but if you do, you may sometimes get errors if the micro-browser tries to interpret a punctuation mark as part of the variable name. For this reason, it is common practice to put the variable name inside brackets. The compiler then substitutes the contents of the variable name first, before placing it in any surrounding punctuation.

In this example we are using **onenterforward** to set the variable "var_name" to the value of "Mr King." This means that when the card is entered from the normal forward direction, the variable var_name will be set. If, however, the card is entered *backwards*, the variable will not be set, but the previous contents will be used. Here are an additional few lines of code that will demonstrate this:

```
<wml>
  <card id="main" title="Show Variable">
    <onevent type="onenterforward">
    <refresh>
    <setvar name="var_name" value="Mr King"/>
    </refresh>
    </onevent>
<p>
Welcome to my application, $(var_name).<br/>
<a href="#card2">Next Card</a>
</p>
</card>
<card id="card2" title="Card 2">
   <onevent type="onenterforward">
     <refresh>
       <setvar name="var_name" value="Mr Smith"/>
     </refresh>
   </onevent>
   <do type="prev">
     <prev/>
   </do>
<p>The name has been changed to $(var_name).</p>
</card>
</wml>
```

When this deck is run, the first card will appear as before. However, if you follow the link to the second card and then press the Back soft key, you will see that the name has been changed to "Mr Smith," as shown here:

Other Ways of Setting Variables

The **setvar** tag is not the only way to set a variable. In fact, until we start using databases, the **setvar** tag is a very limited way of setting variables. After all, if you already know what the text is going to be, you may as well enter the text directly into the code.

Where it starts to become truly useful is when we can assign data that has been entered or selected by users to an internal variable for further use within the application. We can do this by using a **select** list or by using the **input** tag to collect and retain data from the user.

Let's take a look at some simple uses of this.

```
<wml>
  <template>
     <do type="prev"><prev/></do>
  </template>
    <card id="card1" title="Favorite Fruit">
    <p align="center">
    Pick a fruit:
    <br/>
    <select name="fruit" value="Fruit">
      <option value="Oranges">Oranges</option>
      <option value="Apples">Apples</option>
      <option value="Pears">Pears</option>
      <option value="Strawberries">Strawberries</option>
    </select>
    <a href="#card2">Results</a>
    </p>
  </card>
    <card id="card2" title="Results">
    <p align="center">
```

Chapter 6: Interactivity: Forms and User Input

```
      You picked $(fruit)
    </p>
  </card>
</wml>
```

In this code fragment, the user simply picks a fruit from the list. In the first screen (shown on the left), the user clicks on the Results link and is presented with the screen on the right.

The user's selection is shown on the next card:

What this means to you as a developer is that instead of having to write a different card for each selection that the user might make, you only need to write one card that will handle the different variations from the selection list. This makes development and maintenance of your code much easier.

Another way of getting data from the user is by the use of the **input** tag. Although we are going to cover this in more detail in the next section, I just want to demonstrate the use of the input tag in assigning variables. Here is another simple example:

```
<wml>
  <template>
    <do type="prev"><prev/></do>
  </template>
    <card id="card1" title="Enter Name">
<do type="prev"><noop/></do>
```

```
        <do type="accept" label="Go"><go href="#card2"/></do>
        <p align="center"> What is your name?<br/>
<input name="var_name"/>
<br/>
        </p>
   </card>
      <card id="card2" title="Hi!">
        <p align="center">
          Hi there $(var_name)!
        </p>
   </card>
 </wml>
```

If you enter your name in the text box, as shown in the next illustration on the left, pressing OK gives you a screen with a greeting, as in the illustration on the right.

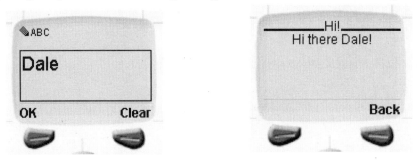

Yes, this is all pretty corny stuff; but these are the basic building blocks of WML, and if you get them down cold, you will be able to build robust applications. If not, you will always be guessing.

THE INPUT TAG

The last major method of getting data from the user is the **input** tag. We briefly touched on this in the previous section to show how to get data from the user and put it directly into a variable, but you will see that it can do far more than that.

As I have already mentioned in some detail, the **input** tag should be used with caution—not because it is difficult or dangerous to use, but because mobile devices are not the easiest input devices in the world. There are, of course, many times when there is no other way to get the data from the user. But the **input** tag should not be used frivolously, as it makes unnecessary work for the user, and this will not be appreciated.

Some things cannot be selected, such as numbers or names, and this is when the **input** tag comes into its own. It is similar to the <input> tag used in HTML, except that it does not have to be used inside a <form> tag with a Submit button.

As we have already seen, the user's data is collected and then assigned to a variable that has been named in the **input** tag through the **name** attribute. You can assign a default value by using the **value** attribute of the **input** tag like this:

```
<input name="var_name" value="The Club"/>
```

This will display on the micro-browser as shown here:

This looks a bit friendlier than the default "[]" that would otherwise appear. And speaking of friendlier, another attribute that makes the **input** tag more explicit is the **title** attribute. Instead of showing the edit box "plain," like so,

we can make it more user friendly by adding the **title** attribute so that the user always knows exactly what item of data it is that she is supposed to be entering. For example, if we use the line

```
<input name="var_name" value="The Club" title="Clubs" />
```

then the title, in this case "Clubs," appears above the edit box, as shown here:

Data Formatting

Although there are no options for actual data validation in WML (there are in WMLScript, but we will fully cover this in the next chapter), we do have some more options on how to control the input while the user is typing it in, and they are covered next.

Size

This is similar to the HTML size attribute and has the same basic specification. It specifies the width in characters of the input data field that the user is permitted to enter data into. However, this attribute appears to be completely ignored by the current WAP devices. You can use the **size** attribute if you want it to be used in the future (without having to modify the code at that time), but it will be ignored for the time being.

Maxlength

You can restrict the user to a specific number of characters to input in the input box. You do this by using the **maxlength** attribute. Please note that it is a very good idea to do this for all input fields, so that you can control the total size that the memory usage will grow to. A typical line of code using this attribute would look like this:

```
<input name="var_name" value="The Club" title="Clubs" maxlength="40" />
```

When inputting the data, users simply cannot enter anything longer than 40 characters.

Type

There are two types that you can assign to the **input** tag. They are **text** and **password**. **Text** is the default and so does not need to be specified, but of course it can be. If I haven't managed to convince you earlier in this book that you should never use the **password** type, go ahead and use it. But please bear in mind how frustrating it can be for the user.

If you do want to use the **password** type, the following line of code demonstrates how it should be used.

```
<input name="var_name" type="password" title="Password" maxlength="8" />
```

The resultant screen will look like this:

Even though the characters are replaced by asterisks, the variable is assigned the actual characters that the user types.

In the previous example, I used "mobile" as the password. If you then display the variable on another screen with the code

```
You typed $(var_name)!
```

you will get the following screen:

Emptyok

This attribute allows you to permit users to continue through the application even if they enter nothing in a data field. The default is False, which means that users cannot continue unless some data is added to the **input** variable.

You might use the **emptyok** attribute if you would like to get some data from the user, but the user might not want to give it if he or she considers the question irrelevant or intrusive. If this is the case, you certainly don't want to bar the user from continuing with the application; so you can use the **emptyok** attribute instead.

The following line of code demonstrates how you might use this.

```
<input name="var_gender" title="Gender" emptyok="true"/>
```

Format

The concept of formatting is simple, but there are a number of variations to be taken into account, which can at first glance make it seem more complex than it really is. The idea is that certain characters act as placeholders in the input edit box, and they dictate which characters or numbers the user can enter. If the character is supposed to be a number, or a date, and the user tries to enter a letter, then the character is ignored.

Here is a simple example of what I mean:

```
<input name="var_age" title="Your Age" format="NN" />
```

A capital *N* means that only a numeric character can be entered. The code "NN" means that two digits are mandatory. If users enter a single digit, they will not be allowed to continue.

Although this method of validation can't trap any errors, such as the user typing **99** instead of a valid month between "01" and "12", at least it can take you part of the way there. This partial validation will at least stop the user from typing a character instead of a number, thus reducing the amount of error checking needed when the data entered is run through any WMLScript validation routines (which we will cover in the next chapter).

Although the use of **format** codes isn't a complete solution by itself, it is free and saves having to take up unnecessary bandwidth by writing code to check for a character, when we already know that there is no way that a character could have been entered.

Several different codes can be used to set the format. I will list them all in Table 6-2, and then we can look at a few examples. You can mix and match any of the codes, with the obvious exceptions of the last two code sequences, as they can only be placed at the end of the format string.

An obvious example would be a telephone number, which could be formatted something like this:

```
<input name="var_phone" title="Telephone Number" format="NNN\-NNN\-NNNN" />
```

When this is run, if you enter the digits **555** into the input box, the dash (-) appears automatically in the fourth character position, as shown below, and then so on until the input string is finished.

Code	Description
A	Any uppercase alphabetical character, or any punctuation mark.
a	Any lowercase alphabetical character, or any punctuation mark.
N	Any single number between 0 and 9.
X	Any uppercase alphabetical character, no punctuation marks.
x	Any lowercase alphabetical character, no punctuation marks.
M	Any character, but the system is expecting an uppercase letter.
m	Any character, but the system is expecting a lowercase letter.
\z	Any character (represented by z here) is mandatory and is entered into the string at this point.
nQ	A given number of characters between 1 and 9, where n is the number and Q is one of the preceding codes. This code sequence can only be used at the end of the string.
*Q	An unlimited number of characters, where Q is one of the preceding codes. Again, this code sequence can only be used at the end of the string.

Table 6-2. WML Format Codes

Another obvious use is for entering dates. The code in the next line will produce a formatted date string.

```
<input name="var_date" title="Enter as MM/DD/YYYY" format="NN\/NN\/NNNN" />
```

Once again, the forward slashes are entered automatically, as shown here:

An example of a mixture of characters and letters that have to be in a given sequence is a social security number, or in the UK, a National Insurance number. A coded sequence for this format string would be

```
<input name="var_ni" title="Enter your NI Number" format="AA\-NN\-NN\-NN\-A" />
```

which will only allow the following kind of output:

As you can see, the **format** attribute is a very good way of getting the data that you want in the way you want it. Used consistently, it makes life easier for both the user and for yourself.

You see? I said it wasn't that hard, didn't I.

SUMMARY

In this chapter, we have covered the basic ways of getting data from the user. Although some of these methods would probably never be used by themselves, when they are used in combination with each other, and with further validation using WMLScript (which we will cover in the next chapter), there is virtually nothing you cannot do to get valid data from the user.

CHAPTER 7

Adding Functionality with WMLScript

A site made of plain WML is fine, and roughly compares to a web site made with HTML. I say roughly, because with WML you will be missing background color and images that can make an otherwise dry web site into an enticing visual feast, even as a set of static pages. You can combine these basic tags with your imagination and create a site that will function beautifully.

But how can you make an outstanding site that really shines for the user? If you are not just going to provide straight static data and have the user be satisfied with that, what else can you provide to make the user's life easier and to provide that "feel good" factor that draws users back?

This is where WMLScript comes into the scene. JavaScript adds a whole new level of functionality for HTML web pages. WMLScript does the same for WML, allowing us to add functionality that is simply not possible with WML alone.

A minor (but vital) example is the ability to add full data validation to user input as it is entered, and to allow the user to take a specific path through the application based on the data that he or she has entered so far. More visible examples are games, calculations, mortgage calculators, interest payments, and so on.

You can achieve yet another level of functionality when you integrate data from a database into your pages. This allows you to work with live data on-the-fly, as the data in the underlying database changes.

We will cover simple database integration techniques in the next chapter. This chapter will introduce you to WMLScript.

WHAT IS WMLSCRIPT?

WMLScript is a *client-side* scripting language. It is very much like JavaScript, which is used on dynamic HTML (DHTML) web sites, but WMLScript has been modified to fit the currently severe limitations of the WAP environment and is in many ways restricted compared to JavaScript.

Client-side means that the code actually runs on the micro-browser itself. The code is downloaded to the WAP device as part of the deck, so that when the code is executed, it does not have to make a trip back to the server to find out what to do next. Everything the application needs is right there. Client-side functionality is therefore much faster in the user's perception.

There are a number of benefits of using WMLScript:

- ▼ **Improves browsing and presentation** WMLScript can enhance the browsing experience and the presentation of WML pages to the user. It can even change the behavior of the micro-browser dynamically. For example, you could use this ability to change the presentation of your pages depending on the type or model of mobile device the user is using. Or you could change an image throughout the application according to the entered gender of the user.

- **Supports advanced user-interface functions**   WMLScript can be used to build and support advanced user-interface functions with full and correct data validation. This is going to depend a lot on the date format that is familiar to the people using the site.
- **Adds intelligence to the client**   You can use WMLScript to provide a degree of perceived intelligence to the client micro-browser and to branch to different sections of the application dynamically, depending on what the user has entered.
- **Provides access to a device's peripheral functionality**   Another major benefit of WMLScript is that it can provide access to a particular device's functionality. By gaining direct access to a mobile phone's memory and ancillary functions, you can add all sorts of clever functionality involving the device itself, all controlled from the micro-browser. This, of course, requires that you know exactly which device you are dealing with. (You can interrogate the mobile device in order to get this information.)
- ▲ **Developers do not need to learn new concepts**   Because the principles of WMLScript are based on existing programming and scripting languages, developers do not need to learn new concepts in order to be able to write it. In fact, developers can pretty much leverage what they already know, and, in some cases, even port existing JavaScript directly. The only clever part in using WMLScript is integrating these scripts into the WML presentation layer. Any understanding or previous experience with programming will help with WMLScript programming. In particular, any use of JavaScript, VBScript or a similar scripting language like Perl would be of benefit, as these are similar in construction and syntax to WMLScript.

THE RULES OF WMLSCRIPT

So, how do we create WMLScript code? Where do we put it? How do we call it from our WML cards?

The first thing you need to know about WMLScript is that the WMLScript code itself is not included within the WML document. In HTML and JavaScript, you include the JavaScript as part of the HTML document. This is *not* the case with WMLScript and WML decks and cards. The WMLScript functions are placed in a separate text file on the server with a file extension of .wmls.

WMLScript uses no main programs, sections, segments, or any other kind of code division that you may think of. You simply write each function as a function. You can have as many functions as you like within a WMLScript file, as long as you do not go beyond the limit of 1,200 characters. You call a WMLScript function from a WML deck and pass parameters directly to the function.

You call a WMLScript function in its external text file in exactly the same way you call a card in a separate deck. The only difference is that the fragment (#) address is the function name, not the name of a card.

To illustrate, you would call a function called "function_one" located in a WMLScript file called "functions.wmls" by using the following statement:

```
<go href="functions.wmls#function_one">
```

Case Sensitivity

WMLScript is a case-sensitive language. All keywords, variables, and function names must use the same case each time in order to be recognized; otherwise you will introduce all sorts of strange bugs into your code.

For example, "dale", "Dale", and "DALE" are all considered to be different items by WMLScript. In order to encourage some consistency, there are several "accepted" conventions of how and when to capitalize your words, but some people like to write in all lowercase, regardless. Others insist that the first letter of each whole word should be capitalized, like "GetLetter", or "InsertSpace". In Java, the convention is to have the first word lowercased and the second word capitalized, as in "getLetter" and "insertSpace".

You should use whichever convention you are comfortable with, but for your own sake, be consistent in whichever convention you choose.

Whitespace and Line Breaks

WMLScript ignores any spaces, tabs, and carriage returns that appear in your script, except those that are included in string literals.

The WMLScript compiler recognizes the three following strings as different:

```
"The cat sat on the mat."
"Thecatsatonthemat"
"Thecatsat    onthemat"
```

The WMLScript compiler recognizes the following three commands as being the same:

```
function helloworld()
function         helloworld()
function
     helloworld()
```

Comments

By now I am sure you don't need me telling you that you should always comment your code as much as possible so that you can figure out what you meant when you wrote it. Or so that somebody *else* can figure out what you meant when you wrote it. You only have to try to sort out somebody else's spaghetti code once in order to vow that you will always comment your code properly.

There are two different types of comments you can use in WMLScript: line comments and block comments.

Line comments are comments that are added as a single line in your code. The line is started with a two slashes (//) and ends with the carriage return at the end of the line. The two slashes can also be added after a working line of code to clarify a particular point. For example:

```
// The temperature in degrees Fahrenheit where this book will ignite
X="451"
```

or

```
X = "451"   // temp in degrees Fahrenheit where this book will ignite
```

A block comment runs across multiple lines and is useful for laying down full descriptions of how a function works, or for laying out the logic for a particular algorithm in a section. Block comments start with a slash and an asterisk (/*) and end with the same characters reversed (*/), which now appears to be the de facto standard for block comments in most current languages. For example:

```
/*
Function InsertSpace(string, x)
This function inserts a space into a string at a specified position
Syntax: InsertSpace(string, x);
Parameters:
string = the text string
x = the char position (zero based) where the space is to be inserted

Example:  InsertSpace("thecat",3);
Result: "the cat".
*/
```

A general principle is that more comments are better than less. In HTML, this would be a liability, as everything you type as a comment in the HTML file will be downloaded to the client machine, wasting bandwidth and giving slower download times. (I have seen some HTML source code that was at least half internal comments, thereby doubling download time.)

In WMLScript, the amount of comments that you use is not an issue, because WML and WMLScript are compiled before they are downloaded to the client micro-browser. On compilation, all of the unnecessary code (unnecessary to the compiler, that is), such as the comments, white space, and so on, is stripped out completely and only the tokenized code is transmitted. So please do me a personal favor, and use lots of comments. After all, the person who might have to maintain your code years from now might be me!

Statements

A statement in WMLScript is one complete unit or instruction. A statement always ends with a semicolon (;). There are no exceptions to this, ever. A lot of the instant bugs that you will get when you first try to run your code will be because a semicolon was omitted from the end of a statement, or added in the wrong place. (As you will see later, what is or is not a full statement in the control constructs, such as if-then-else can be a little confusing at first.)

Here are some valid statements:

```
x="451";
a = b + c;
return a;
```

Code Blocks

A code block is exactly what it sounds like. The block is denoted by the use of braces at the beginning and end (see example below). The code inside the block is to be considered as one indivisible unit. Aside from being mandatory in some places, like function bodies, it is also most useful for avoiding confusion when you have a lot of code in one section.

Blocks can be used to separate different sections of the code for readability, like this:

```
// Initialize all variables for this function
{
    a = "Dale";
    b =  50;
    c = b/2;
}
```

Blocks are most often used in the if statement, where you want to specify several different statements that must be executed if a condition is true, like so:

```
if (username == "Dale")
{
    age = 29;
    personalitytype = "niceguy";
    employable = "yes";
}
```

As you can see, in this case, if the condition is met (if the username is equal to "Dale"), then each one of the following statements will be executed without omissions, as one discrete block.

VARIABLES

Although I have already demonstrated the use of variables in the preceding examples, let's take a look at exactly what a variable in WMLScript is, and how the data is held. There are many different data types: *integers* (whole numbers), *floating point numbers* (with a decimal part, for use with money, for example), *strings* (any of the letters or characters found in any alphabet, also known as a *character* type), and *logicals* (also known as *Booleans*, which can hold the values "true" or "false"). In some languages, there are many more data types, but they aren't used in WMLScript.

So which of these variable types are we concerned with when writing WMLScript? Deep down, we need to be concerned with whether we are dealing with a number or a character. Are we dealing with the number 1, or the character "1". In 999 cases out of 1,000 this is all we will ever really be concerned with.

If we say:

```
x = 1;
y = 1;
z = x + y;
```

then z will contain 2.

On the other hand, if we say:

```
x = "1";
y = "1";
z = x + y;
```

then z will contain "11" as a character string, not a number (because the two strings are joined together, rather than being added as numbers).

How the numbers and characters are held and represented internally is thankfully completely transparent to us as developers. As long as you are aware of the basic difference between strings and numbers, you won't have any real problems in dealing with data types in WMLScript.

How do we create (declare) a variable? It couldn't really be any simpler—just use the **var** keyword as shown here:

```
var x;
```

NOTE: A variable always has to be declared before it is used for the first time.

Variable Scope

As we saw in the Chapter 6, a variable declared in WML is visible globally, even to applications written by other people. Obviously, this is not a terrifically secure state of affairs, and it could potentially give rise to all sorts of strange and elusive errors when the application is run.

Fortunately, WMLScript gives us a completely different scenario, in that you can control exactly which variables can be seen, and by which pieces of code. This means you can use a variable name within a function without having to worry whether or not the variable name has already been used by another function or section of code in another part of the application. Even if you have used the same variable name, the compiler keeps them isolated from one another when the code is compiled.

A simple example of this is as follows:

```
var x;

x = 32;  // Freezing point in Fahrenheit
   { var x;
     x = 0; // Freezing point in centigrade
      // do something with x, which has the value of 0
   }

  // do something with x, which has the value of  32 here
```

The second declaration of x is only visible within the code block in which it has been declared.

The amount of code in which a variable is visible is called the *scope*, and the amount of time between when the variable is created and destroyed (when it goes out of scope) is known as its *lifetime*.

Incidentally, I have used the variable x here to illustrate the point (and the variable could just as easily have been `interest_rate`), but you shouldn't really use this technique intentionally if you can avoid it. It can get confusing in terms of code readability and can lead to unwitting errors being made.

For instance, if there were 50 or 100 lines of code between the two declarations, you might be misled at some point into thinking that the second x was really the first one, and so try to use it inappropriately. The result can be an elusive bug that might take quite a bit of your valuable time to resolve.

OPERATORS

An *operator* is the technical name for any symbol that we use to indicate that a particular mathematical operation is to be carried out, such as the plus symbol (+), which shows that two numbers are to be added together. There are different types of these operators, and we will look at each in turn.

Assignment Operator

The assignment operator is the simplest of all—it is the equal sign (=). To assign a value to a variable, you simply use it like so:

```
x = 5;
name = "dale";
interest_rate = 1.47;
```

As mentioned previously, a variable always has to be declared before it is used, but we can create and initialize a variable all in one step, like so:

```
var x = 5;
var name = "dale";
var interest_rate = 1.47;
```

Arithmetic Operators

The arithmetic operators are the standard plus (+), minus (-), multiplication (*), and division (/) operators that are in common use. Here are some examples:

```
a = 3 + 2;
b = a - 1;
c = b * 4;
d = c / 2;
```

Division has a bonus, as it can provide one of two different results. If you use the standard division (/) operator as above, the result will be given as a floating-point number. The following statement,

```
x = 9 / 4;
```

will assign x the value of 2.25. However, if you use the keyword "div" instead of the slash (/), the result will be given as an integer, and the remainder is dropped:

```
x = 9 div 4;
```

The preceding statement will give x the value of 2.

There is an additional operator that is the reverse of the div operator, which assigns only the remainder of the integer division and drops the whole number part. This is the modulus operator (%), and it is used like this:

```
x = 10 % 3;
```

This statement will give x the value of 1 (which is the remainder of 10 divided by 3).

Bitwise Operators

There are several operators that are used for bitwise operations—operations that operate directly on the binary bits (ones and zeros) within a byte of information that contains the variable data. While these operators are beyond the scope of this book, they are listed in Table 7-1 for completeness.

Increment and Decrement Operators

There are a few shortcuts you can use to make your life a little easier. You may be familiar with these operators from C or JavaScript, but if you are coming across these for the first time, don't worry. You can use the standard arithmetic operators and move into using these shortcuts as you get more familiar with them.

The shorthand arithmetic operators are shown, along with their regular arithmetic equivalents, in Table 7-2.

Bitwise Operator	Description
&	Bitwise AND; returns a 1 in each bit position if bits of both operands are 1's.
^	Bitwise XOR; returns a 1 in a bit position if bits of one but not both operands is a 1.
|	Bitwise OR; returns a 1 in a bit if bits of either operand are a 1.
~	Bitwise NOT; flips the bits of the operand.
<<	Left shift; shifts its first operand in binary representation the number of bits to the left specified in the second operand, shifting in 0's from the right.
>>	Sign-propagating right shift; shifts the first operand in binary representation the number of bits to the right specified in the second operand, discarding bits shifted off.
>>>	Zero-fill right shift; shifts the first operand in binary representation the number of bits to the right specified in the second operand, discarding bits shifted off, and shifting in 0's from the left.

Table 7-1. WMLScript Bitwise Operators

Shorthand Operator	Example	Regular Arithmetic Equivalent
+=	x += 5;	x = x + 5;
-=	x -= 5;	x = x - 5;
*=	x *= 5;	x = x * 5;
/=	x /= 5;	x = x / 5;
%=	x %= 5;	x = x % 5;
++	++x	x = x + 1
--	--x	x = x - 1

Table 7-2. Shorthand Arithmetic Operators

This may look like a neat way to write shorter code, but remember that the compiled code will be exactly the same, whichever way you write it. If in doubt, write for clarity, not because "it looks really cool."

The last two shortcut operators (++ and --) can cause some very interesting bugs if you are not careful. The ++ operator increments a given variable by 1, and the -- operator decrements a given variable by 1. They are used like this:

```
var interest_rate = 7.0;

++interest_rate;    // interest_rate now contains 8.0
--interest_rate;    // interest_rate is now back at 7.0
```

This can get complicated because you can put the ++ in front of the variable, known as *prefix*, or after the variable, known as *postfix*.

```
++interest_rate;    // Prefix
interest_rate++;    // Postfix
```

They may look the same, and indeed, if they are placed on a line by themselves, the end result will be identical—the variable `interest_rate` will be incremented by 1. The trick is the point at which the variable is incremented.

If you say this,

```
var interest_rate = 7.0;

bankcharges = ++interest_rate;
```

the value of the variable `interest_rate` is incremented *before* it is assigned to the variable `bankcharges`, so that `bankcharges` now holds the value of 8.0.

However if you postfix the increment operator, like this,

```
var interest_rate = 7.0;

bankcharges = interest_rate++;
```

the value of the variable `interest_rate` is incremented *after* the value has been assigned to the variable `bankcharges`, so although at the end of the line of code `interest_rate` will hold the value of 8.0, `bankcharges` will only have the value of 7.0.

Obviously, these are two different results. If the variable `bankcharges` is used further in the application (and one assumes that that was the reason for assigning the value to the variable in the first place), then any calculations based on it will be incorrect, because the bank's intention is, of course, to charge as much as they can get away with.

Another classic place to use these operators is in a control loop, where you simply increment a counter by 1 every time you go around the loop. If you use this counter somewhere inside the loop itself, then you really have to make sure that it has the value that you intended it to have when you use it. A loop counter that counts from 0 to 9 is *not* the same as a loop counter that counts from 1 to 10, even though the loop itself runs the same number of times.

Logical Operators

While the operators that we have looked at so far deal with numerical values, logical operators deal with comparisons between the logical values *true* and *false*. It is the ability to make logical decisions based on different values that allows programmers to have their code make choices that can send the user down the correct branch of the application.

Most normal expressions can be evaluated to be true or false, like this one:

```
X > 3;
```

If x were any number up to 3, then this expression would evaluate to false. If x were any number larger than 3, then this would evaluate to true.

> **NOTE:** If you want check the equality of two sides of an expression, you have to use a comparison operator consisting of two equal signs (==). We will be dealing with these in the next section. If we were to write x = 3, the answer would always be "true", as x would be assigned the value of 3 and so would be equal to 3. Little things like this can make a developer"s life exciting.

The logical operators for WMLScript are listed in Table 7-3.

Logical Operator Symbol	Text	Description
&&	AND	AND returns true only if *both* of the operands are equal to "true". If either of the operands is false, then the whole expression returns false.
||	OR	If either one of the operands is true, then the expression returns true. The expression will only return false if both of the operands are false.
!	NOT	NOT simply returns the opposite of the evaluated expression. It reverses it so that the true becomes a false, or a false becomes a true.

Table 7-3. WMLScript Logical Operators

You will probably find yourself using these logical operators a lot, and particularly in conjunction with the comparison operators which follow next.

Comparison Operators

Comparison operators are different from the operators we have covered so far, because they only produce a logical result of true or false. The comparison operators are listed in Table 7-4.

The exactly-equals operator (==) is the easiest comparison operator to understand, but it is also the one that will have you unwittingly writing bugs into your program. The reason is fairly simple; you will find one day that you have written something like:

```
var x=10;

if (x=3) {
  //do something special
}
```

Operator	Description
==	Exactly equals
!=	Not equal to
>	Greater than
<	Less than
>=	Greater than or equal to
<=	Less than or equal to

Table 7-4. WMLScript Comparison Operators

To the casual observer, this looks fine, and you would think that the "if" control structure will never be run. If you are looking through the code, you can easily miss it altogether. Unfortunately, in this example you are not *comparing* x to the value 3, you are *assigning* the value 3 to x! This means that regardless of what you had in the variable x before, you now have the value 3.

This also means, quite incidentally, that the if control structure will always be executed, because the statement x = 3 will always evaluate to true.

If you had written this,

```
var x=10;

if (x==3) {
  //do something special
}
```

the code would have performed as you would have originally intended—if x is equal to 3, then "do something special."

The rest of the comparison operators are straightforward. Here are some simple examples:

```
var x=10;

if (x != 3) {
  // do something
}
```

The preceding code will run because x is not equal to 10.

```
var x=10;

if (x > 3) {
  // do something
}
```

Again, the preceding code will run because x is greater than 3.

```
var x=10;

if (x < 3) {
  // do something
}
```

The preceding code will not run, as x is not less than 3.

The remaining two operators, >= and <=, are also self explanatory, greater than or equal to, and less than or equal to.

String Concatenation

A special use of the plus operator (+) operator is to join two strings. The following expression will result in the variable x containing "DaleBulbrook":

```
x = "Dale" + "Bulbrook";
```

If I want a space between the two words, I would have to do something like this:

```
x = "Dale" + " "+  "Bulbrook";
```

One point to note here is that if one of the operands in the expression is a string, WMLScript is supposed to attempt to convert any data types that are not string type into strings in order to do the concatenation. I have not found this to be reliable across all simulators and do not trust this to occur correctly. In actual fact, relying on the "shortcut" behavior of any language to solve a programming problem is always dangerous, as you never know when the behavior might be changed.

The Comma

The comma can be used to separate several expressions on one line. However, if you are doing assignments, this does not work in the way that you would expect. For example:

```
tempVar = 2 + 2, 3 + 3;
```

In this instance, you might logically think that tempVar would get the value of 4. However, when using the comma to separate expressions, the assignment is made by the *last* expression. In the above example, tempVar would have the value of 6.

You might therefore think that the comma operator has no value at all, but this is not strictly true. You can write such lines as:

```
tempVar = y++, --z;
```

which will increment `y` by one, and then decrement `z` by one and assign the value to `tempVar`. However, while this will work, it does not improve the readability of the code at all. There is nothing that can be gained by writing the above example in preference to:

```
y++;
tempVar = --z;
```

The typeof Operator

The `typeof` operator is called an operator, although it works more like a function in practice. Although you can't assign a specific data type by name to a variable, you can at least determine what the data type of a variable is at any given point by using `typeof`. The `typeof` operator returns a number that indicates the data type.

A simple example would be as follows:

```
var x = 1.414;    // The square root of 2
var z;

z = typeof x;   // z will contain the value 1, as it is a Float.
```

The return values that represent the different data types are listed in Table 7-5.

In case you are wondering when you might use `typeof`, a good example is when you want to perform different actions depending on the data type that is being held in a variable. Using an `if` control structure (the details of which are coming up soon!), you can do clever things like allowing a user into a restricted area of the application if he or she types a specific floating-point number, or a given integer, instead of an alphabetic password.

Data Type	Value Returned
Integer	0
Float	1
String	2
Boolean	3
Invalid	4

Table 7-5.   Return Values for the typeof Operator

By testing for the data type first, you can then follow a particular branch and test for specific numbers or characters and ignore the other branches of the application altogether. This can reduce overhead and turnaround time quite drastically, as you can keep the sizes of cards and decks small by not downloading or performing unnecessary validation code.

The isvalid Operator

Again, the `isvalid` operator is called an operator, but in practice it is a very simple function that simply tells you whether a variable contains a valid value or an invalid value. It returns true if the value is valid, and false if the value is invalid.

You can use `isvalid` for basic testing before you get into the guts of the application, to make sure that you are only operating with valid data. You can also use it to check for things like "division by zero" errors, which can cause an application to crash destructively.

A simple example is as follows:

```
var x = 100;
var y = 0;
var z;

z = x / y;      // z will contain an "invalid" result

z = isvalid x;  // z will contain the value "false"
```

The Conditional Operator

The conditional operator is a simple but amazingly useful operator that you will probably use regularly for simple tests on data. The syntax is as follows:

```
result_x = condition_x ? expression_1 : expression_2;
```

First of all, `condition_x` is evaluated to see if it is true or false. If it evaluates to true, then `expression_1` is evaluated, and the result of that is then assigned to `result_x`. If the `condition_x` evaluates to false, then `expression_2` is evaluated, and the result is assigned to `result_x`. Fortunately, this is easier to demonstrate than it is to explain.

Here's an example:

```
var overtime_rate;
var day = "Sunday";
var hour_rate = 5;
var hours = 8;
var pay;

overtime_rate = (day = "Sunday") ? 2 : 1;

pay = hours * (hour_rate * overtime_rate);
```

Here, `overtime_rate` is set to a value of 2 (double time) if the day is a Sunday, and otherwise `overtime_rate` is set to 1. The next line sets pay as the number of hours worked (8) multiplied by the hourly rate (5) times the overtime rate (2 or 1, depending on the day of the week). In the case above, the result will be 80.

The use of the parentheses brings us neatly to the idea of operator precedence—the sequence in which the operators are evaluated in an expression—which is discussed in the following section.

Going back to data validation, you could use the conditional operator to force a variable to take a specific value in the event of a division by zero error, like this:

```
var x = 100;
var y = 0;
var z;

z = ( isvalid x/y ) ? x / y : 0;
```

Testing the validity first makes sure that the result, z, will contain a valid number before you move on to the rest of the code. (Of course, the default value could be any other value instead of 0.)

Operator Precedence

Operator precedence—the sequence in which the operators are evaluated in an expression—must be taken into consideration when you are putting any kind of arithmetic functionality into your application.

Consider this simple example:

```
x = 1 + 2 * 3;
y = (1 + 2) * 3;
```

Here, x will be 7, and y will be 9. This is because the rules of operator precedence (which are standard mathematical rules, not just applicable to WMLScript) dictate that multiplication takes place before addition. However, the contents of any brackets are evaluated before any expressions outside of brackets.

In the previous expressions, x is 7 because the multiplication is done first, and then the addition. In the second case, y is 9 because the expression in the brackets is evaluated first.

A scale of operator precedence follows shortly, but you can go cross-eyed trying to figure out how to write an algorithm and make it all work properly. It is more important to keep the following simple rule of thumb in mind: *Always place brackets around the parts of the expressions that you want to have evaluated together as a unit.* As far as operator precedence is concerned, that really is all you will ever need to know in order to keep all of your mathematical equations working perfectly. It often helps to make the code more readable as well.

You could write the earlier expression for x in either of the following formats:

```
x = 1 + 2 * 3;
x = 1 + (2 * 3);
```

Both of these give identical results, but the second version is totally unambiguous. It is *really* clear that 2 is multiplied by 3 first, before adding the 1! If you bear this in mind, you can never really go wrong.

The scale, of operator precedence is as follows, in order of priority

```
++    --   !   ()    typeof    isvalid
*    /   div   %
+    -
<<   >>   >>>
<    <=   >   >=
==   !=
&
^
|
&&
||
?  :
=    +=   -=   *=   /=   div=   %=   <<=   >>=   >>>=   &=   ^=   |=
,
```

CONTROL CONSTRUCTS

Finally, we get to the things that give any programming language its real power and versatility. A control construct allows the programmer to make decisions based on the value or values of data entered, and have the program take a different route depending on those values. The control constructs in WML are IF...ELSE, WHILE, and FOR. We will examine these one at a time.

If Statements

An `if` statement consists of the evaluation of a condition followed by one or two separate statements or blocks of statements. If the evaluated value is true, the first statement or block of statements is executed. If the evaluated condition is false, then the second statement or block of statements (if it exists) is executed.

Here are some examples that will also demonstrate the syntax:

```
var x = 1;
```

```
if (x == 1) {
   rate = 6.5;
}
```

This is the simplest form of the `if` statement, where there is just one statement to be executed if a simple condition evaluates to true. The condition to be evaluated follows the `if` statement and is always enclosed in round brackets.

A code block is then opened, and the statements to be executed if the condition evaluates to true are placed inside it. In this case, the variable `rate` is assigned a value of 6.5, but the statements can be any valid WMLScript statements, including other `if` statements.

Notice that I have used the `==` operator to test for equality. As mentioned earlier in this chapter, if you use the single equal sign (=), then an assignment will be made and the condition will always be true.

Here's another example:

```
var age = 21;
var msg;
var nextdeck;

if (age >= 18) {
   msg = "Please place your vote";
   nextdeck = "accept.wml";
}
else {
   msg = "Sorry, you're too young to vote";
   nextdeck = "reject.wml";
}
```

Here, the variable `age` is set to 21. When the code is run, the condition `age >= 18` evaluates to true, and the first code block is run. The variable `msg` is set to Please place your vote, and the variable `nextdeck` is set to accept.wml.

You can also nest `if` statements within other `if` statements, like so:

```
if ( age >= 18 ) {

   if ( age >= 65 ) {
      msg = "Welcome to Retirement Plaza";
   }
   else {
      msg = "Welcome to Homes of the Future";
   }
}
else {
   msg = "You are too young to buy real estate";
}
```

This code checks not only for the age being greater than or equal to 18, but also then checks that the age is not more than 64. The message is amended to suit each age group based on the value in the age variable.

While Statements

The while and for statements both create *loops* (for statements are explained in the next section). In a loop, an expression is evaluated and, if it is true, a statement or block of statements is executed. The expression is then evaluated again, and if it is still true, the statement or block of statements is executed again. This process continues until the expression evaluates to false, at which time the loop exits and the code following the loop is executed.

Here is a simple example of a while loop:

```
var counter = 0;
var total = 0;

while (counter < 5) {
    counter = counter + 1;
    total = total + counter;
};
```

After this code has run, the variable total will contain the value 15. The while loop repeats for as long as the specified condition is true; in this case, as long as counter < 5 is true.

Note that in a while loop the condition is evaluated at the beginning of the loop, so if the condition is not met the first time, the statements in the loop will never run.

For Statements

The for loop is different from a while loop. The for loop has a start condition and an end condition which are both set at the beginning of the loop cycle. The loop will then execute as long as the end condition has not been met.

Here is an example of a for statement:

```
for (var index = 1; index <= 10; index++) {
    anotherFunction(index);
};
```

As you can see, the syntax of the for statement is quite different from anything we have encountered so far. The for statement consists of three optional expressions enclosed in parentheses and separated by semicolons, each with different functions as shown here:

```
for (initialization statement; condition statement; increment/decrement statement) {
```

This is followed by a statement or block of statements to be executed, known as the body of the loop.

The first expression (`var index = 1`, in the example) is used to initialize a counter variable, which is then used to control the number of times the loop is executed. This expression can declare a new variable with the `var` keyword. The scope of the declared variable is the rest of the function.

The second expression (`index <= 10`, in the example) can be any expression that evaluates to a Boolean (true or false) value. This condition is evaluated on each execution of the loop. If the condition is true, the loop body is performed. Including this conditional test is optional, but if it is omitted, the condition always evaluates to true. You might wonder why you would ever need to have an infinite loop (one that never ends), but there are occasions where this is useful. You might want to have a menu that is always displayed, with `if` statements that pick up the changing inputs of the user.

The third expression (`index++`, in the example) is generally used to update or increment/decrement the counter variable. This statement is executed as long as the condition is true.

In the previous example, the body of the loop simply consists of a call to `anotherFunction()`. In that example, it will be called ten times, and each time it will be passed the parameter `index`, which will be the numbers from 1 to 10 in turn.

You can control the amount by which the ""counter variable is incremented by changing the third expression in the `for` statement, like so:

```
for (var index = 0; index <= 20; index += 2) {
    // statements
}
```

In this example, `index` will be incremented by 2 on each loop, so the values of `index` would be 0, 2, 4, 6, ... 20 on successive iterations.

Stopping Loops or Skipping Unnecessary Loop Statements

You will come across situations where you want to stop the loop (whether it is a `for` or `while` loop) from continuing. For instance, if you are half way through a loop, and the function you are calling returns an error, you don't want to necessarily go on to call the function repeatedly until the loop finishes. Similarly, you may sometimes want to skip the remaining statements in the loop for a particular value of the counter, but you do want to increment the counter and run the loop the remaining number of times.

There are two statements that you can use to perform these actions: `break` and `continue`.

Break Statement

The `break` statement is used to terminate the current `while` or `for` loop and continue the program execution from the statement immediately following the terminated loop.

Here is an example of the `break` statement in action.

```
for (var index = 1; index <= 10; index++) {
    if (anotherFunction(index) == "Error") {
        break;
    }
};
```

Here, we are combining two actions in one line by calling the function anotherIndex(index) as part of the if statement. If the function returns the value Error, then we no longer want to continue the loop, and the break statement is called.

Incidentally, we could do the exact same thing in a more readable way by writing the loop like this:

```
for (var index = 1; index <= 10; index++) {
    var test;
    test = anotherFunction(index);
    if ( test == "Error" ) {
        break;
    }
};
```

Continue Statement

The continue statement is used to terminate the current execution of a while or for loop and continue the execution of the loop with the next iteration.

Here is an example using the continue statement:

```
for (var index = 1; index <= 10; index++) {
    var test;
    test = anotherFunction(index);
    if ( test == "Error" ) {
        continue;
    }
    flag = "success";
      // more statements
};
```

In this example, any error will cause the continue statement to be reached, and the loop will jump to the beginning of the loop. The remaining statements, (flag = success, and so on) will not be executed on that iteration of the loop.

Remember that the continue statement does *not* terminate the execution of the loop. If it is executed in a while loop, execution returns to the while condition, and if it is executed in a for loop, execution returns to the update expression.

Finally, you should note that you are only allowed to use a break or a continue statement within a for or while loop.

RESERVED WORDS

Before we get into building functions with what we have learned so far, I will give you a list of words that are reserved by WMLScript for the language and syntax construction. If you skim over this list now, you won't be surprised when you are coding and you get an error because you wrote something like `var while = 0;`.

In WMLScript the current reserved words are shown in Table 7-6. This is the current list as I write, but this list can change with new versions of WMLScript. To avoid future problems, it is best to create your own naming convention that is not likely to conflict with any reserved words, past, present, or future, and stick with it.

For example, one convention that works well because it kills two birds with one stone is to prefix your variable name with a character that indicates the data type that the variable is supposed to hold.

```
var cName;      // Name, should be character type (string)
var iAge;       // Age, should be an integer
var dInterest;  // Interest rate, should be a floating point number
```

With a system like this, you can look at a variable that is called iAge when you are debugging, or when you come back to the code in a year's time or more, and if it has a value of "12.56", you know immediately that this value is wrong.

access	agent	break	case
catch	class	const	continue
debugger	default	delete	div
div=	do	domain	else
enum	equiv	export	extends
extern	finally	for	function
header	http	if	import
in	isvalid	lib	meta
name	new	null	path
private	public	return	sizeof
struct	super	switch	this
throw	try	typeof	url
use	user	var	void
while	with		

Table 7-6. Current Reserved Words for WMLScript

NOTE: This is the convention I prefer to use if I have a choice, but consultants often have to use conventions that have already been laid down. One system I worked on recently had all of the hundreds of variables named with six-digit numbers, prefixed with the letter "K". Everyone needed a manual two inches thick to look them up, and without the manual, we were quite literally lost. Easily 20 percent of the build time was spent looking up these anonymous variables. Such a waste.

FUNCTIONS

A *function* is a part of a program that performs a specific task, and that can be called (started running) from more than one place within the program as many times as necessary. A function can also have input parameters and output parameters.

We have already seen snippets of functions in the examples of WMLScript we have used so far, but we have not as yet looked a full function or seen how it works. A function must be coded in a file separate from the WML deck, and the filename takes an extension of .wmls.

The structure of any function is quite simple:

```
extern function functionname(parameter1, parameter2, etc.) {
    // function body (WMLScript)
return resultvalue;
}
```

The `extern` keyword is optional, but it is needed if you want to call the function from an external file (like a WML deck). If this keyword is left off the function declaration, then the function can only be called by other functions within the same WMLScript file.

Keeping functions private this way is not necessarily a bad thing at all. You may have several internal functions that you don't want exposed to the rest of the world, or that might really mess up the internal logic and procedures of the rest of the application if they were accidentally called out of sequence.

The `return` statement is also optional, and as you will soon see, is not used very much when the function returns values to a WML page. Instead, we will use WMLScript library functions.

If no `return` statement is specified, or if for some reason none of the function `return` statements are executed, the function returns an empty string as a default.

Parameters

Parameters are also known as *arguments,* and these are the values that are passed to a function in order to provide it with the values it needs to work on to produce the required result. They appear in the function declaration like so:

```
extern function checknum(t_displayfield, t_num) {
    // check the number, etc.
}
```

In this example, the variables being received, `t_displayfield` and `t_num`, are automatically declared as being variables by virtue of appearing in the function declaration. They do not have to be declared again with the `var` keyword.

When the function is called, the number of parameters has to match in both the calling and the called code. In this case, the function expects to receive two parameters, so the calling code must send two. Sending one, or three, or no parameters will produce a fatal error. The previous example function could be called with a statement like this:

```
<a href="functions.wmls#checknum(disp_num, inp_num)">Check Number</a>
```

Bear in mind that the data type of the parameters you send is not automatically checked as the number of parameters is. If the function is expecting two numbers, and you send two character strings, you are expected to do any necessary data checking and validation within the function before you do anything else with them. If you send the wrong number of parameters, the function will not even begin to execute, which is quite a different thing altogether.

Calling Functions

Different programming languages call functions in different ways. Usually the function is just called where you want to use the result directly. In HTML, you can place a JavaScript function directly into the HTML and have it display whatever or however you like. This is the one place where WMLScript is quite different from the usual procedures that developers have come to know.

As mentioned earlier, you cannot have WMLScript in the WML deck. WMLScript has to live outside the WML file in a separate file with a .wmls extension.

If you think about it, the WML deck has already been tokenized by the WAP Gateway, and the variables have already been declared and possibly already assigned values before it was sent to the micro-browser. So if the variable is assigned a different value by the function, then the function has to physically refresh the display of the card in the micro-browser for the changes to be viewable by the user.

These actions are done by using the WMLBrowser library functions, `WMLBrowser.setvar()`, and `WMLBrowser.refresh()`. Let's take a look at this in action, and see how it works.

First, the WML deck:

```
<wml>

<card id="card1" title="Calculate the Cube" newcontext="true">

<do type="accept" label="Cube Value">
    <go href="functions.wmls#cube('result',$(number))"/>
```

```
</do>

<p>
<!-- Get the user input -->
    Number: <input type="text" name="number" title="Number:"/>
<br/>

<!-- Display the result -->
    Cube Result: <u>$(result)</u>
</p>
</card>
</wml>
```

Next the function, in a file called functions.wmls:

```
extern function cube(varName, number) {
var result;

result = Float.pow(number, 3);

WMLBrowser.setVar(varName, result);
WMLBrowser.refresh();
}
```

Let's look at this in a little more detail. First, the deck is set up so that the user can enter a number. A placeholder variable (called `result`) is set up so that we have somewhere to put the result for display back to the user. Here is the relevant section of the deck code:

```
<p>
<!-- Get the user input -->
    Number: <input type="text" name="number" title="Number:"/>
<br/>

<!-- Display the result -->
    Cube Result: <u>$(result)</u>
</p>
```

We then create the do element that will perform the actual function itself, and put this into the top of the deck. (This looks like we are writing the code backwards, but it can be much easier to write code in logical sections and then insert it where it needs to go in the deck, rather than try to write all of the code "from the top down".)

```
<do type="accept" label="Cube Value">
    <go href="functions.wmls#cube(''result'',$(number))"/>
</do>
```

You will notice that we are passing two parameters: `result`, which is the result placeholder, and `$(number)`, which is the variable containing the number that has been entered by the user. When the user enters a number in the number variable via the `input` statement, and then presses the soft key labeled Cube Value, the function is called.

Let's say that the user enters the value "6". The function `cube()` is called like so:

```
functions.wmls#cube('result', 6)
```

The function, once again, looks like this:

```
extern function cube(varName, number) {
var result;

result = Float.pow(number, 3);

WMLBrowser.setVar(varName, result);
WMLBrowser.refresh();
}
```

First of all, a *local* variable called `result` is created. The `number` is calculated to the power of 3 with a standard library function, `Float.pow(number, 3)`. (We will cover library functions in more detail in the next section of the chapter. For now all you need to know is that a lot of the hard work of creating functions that do basic tasks has already been done, and these are at your command.)

Finally, the *placeholder* variable `result`, which was passed to the function and is referred to locally within the function as `varName`, is set with the value of the *local* variable called `result`, and the last line of the function instructs the browser to refresh the display of the card. Because the value of the variable `result` within the card has been changed by the function, the display now shows the new value—in this case, "216", which is 6 cubed.

The screen displays for this example are shown in the following six illustrations:

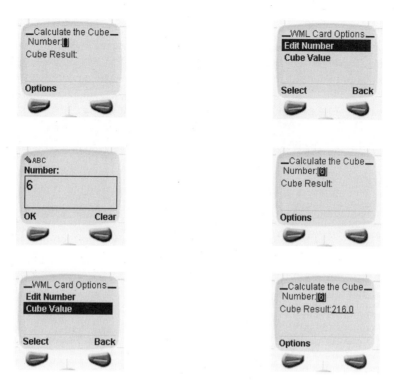

THE STANDARD LIBRARIES

The *standard libraries* are collections of functions that are contained within the micro-browser itself. One major advantage of this, of course, is that you can avoid a round trip to the server whenever you use one of these functions. The second major advantage is that you don't have to reinvent the wheel when you want to do something basic, like extract a substring from a string.

WMLScript provides six libraries that you can use: Dialogs, Float, Lang, String, URL, and WMLBrowser. The functions in these libraries are called just like any other function you may have written, but you also must put the name of the function library first, followed by a period or full stop:

```
WMLBrowser.setVar(varName, result);
WMLBrowser.refresh();
```

This simply means "the setVar() function belonging to the WMLBrowser library," or "the refresh() function belonging to the WMLBrowser library." If you omit the library name, the compiler will look for a user function of the same name.

Some library functions are more directly useful to developers than others, and part of the learning curve is finding out which functions are the useful ones, and which ones you just need to be aware of in case you happen to need them at some point. I can tell you right now that you will use `WMLBrowser.setVar()` and `WMLBrowser.refresh()` all of the time, and the average developer will use something like `URL.getScheme()` (which returns the protocol section of a URL string) hardly at all, if ever.

The point is that you do not need to know all of these functions and libraries by heart right away. The best thing to do is skim through the library functions and their descriptions so that you get a feel for what can be done with them. Then, when you are writing code and you realize that you want to be able to extract the two-digit month from a date string, you will remember that there is a function that can do this for you (`String.subString`), and you can go and look it up as you need it.

The following sections introduce the six libraries and the functions within them, without getting into examples of each one, so you have an overview of what they are and what they do.

The Dialogs Library

There will be many times where you will want to either alert the user to something, get the user to confirm something ("Do you REALLY want to delete that?"), or prompt the user for some information on-the-fly, like a telephone number to dial for example. These are called Dialogs, and the functions for the Dialogs library are listed in Table 7-7.

The Float Library

This library is not to do with waterproofing your application—this library contains functions for dealing with floating-point numbers and for converting them to and from the basic string variables that are used in WMLScript. The float library functions are shown in Table 7-8.

Function	Description
`Dialogs.alert`	Displays a warning message to the user, and then returns them to the previous card.
`Dialogs.confirm`	Prompts the user for a confirmation or cancellation on an asked question.
`Dialogs.prompt`	Shows the user an input box, and gets them to fill it in.

Table 7-7. Dialogs Library Functions

Function	Description
`Float.ceil(number1)`	Returns the integer value closest to, but not less than, `number1`.
`Float.floor(number1)`	Returns the integer value closest to, but not greater than, `number1`.
`Float.int(number)`	Returns the integer part of the parameter.
`Float.maxfloat()`	Returns the maximum floating-point value supported by this browser.
`Float.minfloat()`	Returns the minimum floating-point value supported by this browser.
`Float.pow(number1,number2)`	Returns the value of `number1` raised to the power of `number2`.
`Float.round(number)`	Returns the value of `number` rounded to the closest integer value.
`Float.sqrt(number)`	Returns the square root of `number`.

Table 7-8. Float Library Functions

The Lang Library

The Lang library contains the core WMLScript functions that are either basic to the correct functioning of other library functions, or that don't quite fit into other library categories. Lang in this case means basic to the language (WMLScript). Table 7-9 lists the Lang library functions.

Function	Description
`Lang.abort(errorDescription)`	Aborts the WMLScript and returns `errorDescription` to the calling function.
`Lang.abs(number)`	Returns the absolute value of `number`.

Table 7-9. Lang Library Functions

Function	Description
`Lang.characterSet()`	Returns an integer representing the MIBenum (this is a unique value for use in Management Information Base documents to identify a particular character set) of the character set supported by the WMLScript interpreter currently being used.
`Lang.exit(value)`	Exits the WMLScript and returns the `value` specified to the calling function.
`Lang.float()`	Returns true if floating-point values are supported by this micro-browser.
`Lang.isFloat(string)`	Returns true if `string` can be successfully converted to a floating-point value.
`Lang.isInt(string)`	Returns true if `string` can be successfully converted to an integer.
`Lang.max(number1, number2)`	Returns the largest of `number1` and `number2`, or returns `number1` if they are both equal.
`Lang.maxInt()`	Returns the maximum supported integer value of this micro-browser.
`Lang.min(number1, number2)`	Returns the smallest of `number1` and `number2`, or returns `number1` if they are both equal.
`Lang.minInt()`	Returns the minimum supported integer value of this micro-browser.
`Lang.parseInt(string)`	Returns the integer value of `string`.
`Lang.parseFloat(string)`	Returns the floating-point value of `string`.
`Lang.random(integer)`	Returns a random integer between 0 and `integer`.
`Lang.seed(integer)`	Seeds the random number generator with the value of `integer`.

Table 7-9. Lang Library Functions *(continued)*

The String Library

The String library contains all of the functions necessary for you to be able to manipulate character strings of any size. This is one of the libraries that you will be using a lot if you are dealing with any kind of character or string input or display, and particularly if you are working with or handling arrays. Arrays in WMLScript are always composed of strings, and five of the functions in this library are ideally suited to handling array elements. Table 7-10 lists the String library functions.

Function	Description
`String.charAt(string, index)`	Returns a string containing the character at the position `index` in `string`.
`String.compare(string1, string2)`	Compares `string1` with `string2` and returns –1 if `string1` is less than `string2`, 0 if they are equal, or 1 if `string1` is greater than `string2`.
`String.elements(string, character)`	Returns the number of elements present in `string` separated by `character`.
`String.elementAt(string, index, character)`	Returns the `index`"th element in `string` separated by `character`.
`String.find(string, substring)`	Returns the position where `substring` is found in `string`. The value –1 is returned if there is no match.
`String.format(format, value)`	Returns the formatted result attained by using the format value. `value` can be one of several things, depending on the type of output you want - string, integer, float and so on.
`String.insertAt(string, element, index, character)`	Returns `string` after `element` has been inserted in the `index`"th element separated by the specified `character`.
`String.length(string)`	Returns the length of `string`.
`String.isEmpty(string)`	Returns true if `string` is empty, false otherwise.

Table 7-10. String Library Functions

Function	Description
`String.removeAt(string, index, character)`	Returns `string` after the `index`"th element separated by `character` has been removed.
`String.replace(string, oldstring, newstring)`	Replaces all occurrences of `oldstring` in `string` with `newstring`.
`String.replaceAt(string, element, index, character)`	Returns `string` after the `index`"th element separated by `character` has been replaced by `element`.
`String.squeeze(string)`	Returns `string` after all whitespace inside the string has been removed.
`String.subString(string, start, length)`	Returns a string containing all characters in `string` from position `start`, `length` characters long.
`String.toString(number)`	Returns `number` converted to a string.
`String.trim(string)`	Returns `string` with all leading and trailing spaces removed.
`String.toString(value)`	Returns `value` converted to a string.

Table 7-10. String Library Functions *(continued)*

The URL Library

The URL library doesn't contain a library of URLs, but instead it contains functions that can validate and manipulate URL strings. Table 7-11 shows the URL library functions.

Function	Description
`URL.escapeString(string)`	Returns `string` "escaped" to convert any special characters to the hex equivalent.
`URL.getBase()`	Returns the URL the script was called from, but without the function fragment attached.

Table 7-11. URL Library Functions

Function	Description
URL.getFragment(URLString)	Returns the fragment part of URLString only.
URL.getHost(URLString)	Returns the domain name part of URLString only.
URL.getParameters(URLString)	Returns any parameter part of URLString only.
URL.getPath(URLString)	Returns the path below the domain name in URLString, which is to say the location of the card on the current host.
URL.getPort(URLString)	Returns the port being accessed by URLString. If present, the port number is found at the end of the domain name and separated from it by a colon, e.g., http://www.webdesigns.ltd.uk:8080/
URL.getQuery(URLString)	Returns the query parameters from URLString.
URL.getReferer()	Returns the relative URL of the resource that called the WMLScript.
URL.getScheme(URLString)	Returns the scheme, or protocol, from the front of URLString.
URL.isValid(URLString)	Returns true if URLString is in a valid format, or false otherwise.
URL.loadString(URLString, document type)	Returns the full contents of a file given at URLString, provided the document type matches the second parameter.
URL.resolve(base URLString, relative URLString)	Returns a complete URL string constructed by adding the relative part to the base part.
URL.unescapeString(string)	Returns string "unescaped" to change any converted hex characters back to the special character equivalent.

Table 7-11. URL Library Functions *(continued)*

The WMLBrowser Library

The WMLBrowser library contains functions by which WMLScript can access and control the WML micro-browser. You will be using some of these a lot. Table 7-12 lists the WMLBrowser library functions.

ARRAYS

Arrays, in my opinion, are probably one of the most useful and underrated of all of the developers tools. Arrays in WMLScript are always represented as strings. In the String library, you can use five of the String functions to manipulate array elements. These functions, `elementAt()`, `elements()`, `insertAt()`, `removeAt()`, and `replaceAt()` can be used to do virtually anything you like with a two-dimensional array. You can even use these functions as building blocks to build your own library of advanced array handling functions.

Function	Description
`WMLBrowser.getCurrentCard()`	Returns a string containing a relative URL to the card being displayed by the micro-browser.
`WMLBrowser.getVar(variable)`	Returns the value of `variable`, or an empty string if `variable` does not exist.
`WMLBrowser.go(URLString)`	Sends the browser to the location in `URLString` when the current script has finished executing.
`WMLBrowser.newcontext()`	Clears the browser history and all existing WML variables.
`WMLBrowser.prev()`	Returns the browser to the previous card in the history stack when control is returned to the browser.
`WMLBrowser.refresh()`	Refreshes the display of the card in the browser when control is returned to the browser.
`WMLBrowser.setVar(variable, value)`	Returns true if `variable` is successfully set to the new `value`, or returns false if it fails.

Table 7-12.   WMLBrowser Library Functions

Arrays are particularly useful for repetitive processing and for conserving space. If you have five different options, say for a menu, you can create an array containing the options, and then just loop through the array, placing each option on the browser display.

Instead of hard-coding the options in the deck, you can change the options or the number of options simply by changing the array in the WMLScript file, without having to change any other code at all. If you want different actions to be taken for each array option, you can create two arrays whose index elements match. For example, action number 3 matches question number 3, and so on.

Because you can use any separator character that you like between elements, you can even get cleverer and use something like "question, response $ question, response". By splitting the elements along the dollar ($) separator, you then have the question and response pair in one array element, which you can then split up into its separate parts by using the comma (,) separator. In this way, all of the questions and responses are in one array, and you can instantly see that you have the correct matched response to a given question. An example of this would be as follows:

```
("Founder of the Ford Motor Company", "Henry Ford" $ "First British
Female Prime Minister", "Margaret Thatcher")
```

String indexes start at 0—the first character in a string is at position 0. This is one of those *important* points that you should engrave into your forebrain with lasers. If you have an array with 10 items in it, and you write a loop to list them that looks like the following, you will get into trouble:

```
var temp_array = ("1", "2", "3", "4", "5", "6", "7", "8", "9", "0");
var temp_value;

for (x=1 ; x<=10; x++) {
   temp_value = String.elementAt(temp_array, x, ",");
   // Now do something with temp_value
}
```

This loop will run using the value "2" as the first element, instead of the "1" that you intended. To get the first element, you have to write the `for` statement like this:

```
for (x=0 ; x<=9; x++) {
   temp_value = String.elementAt(temp_array, x, ",");
   // Now do something with temp_value
}
```

Similarly, you might find that you specifically want the fifth element in the array. You would then have to ask for `element(4)`, as the sequence runs 0, 1, 2, 3, 4. But don't worry. Once the "laser etching" has taken hold, this method of counting will become second nature, and indexes and arrays will hold no fear for you.

Now that we have covered the basic concepts, the syntax of the array-handling functions is shown in Table 7-13.

Function	Description
`String.elements(array, separator)`	This function returns the number of elements in `array`, if we use the `separator` specified. This is useful if you have a dynamic array, or if you want to write a generic array-handling function.
`String.elementAt(array, position, separator)`	This function returns the value of the element at the specified (0-based) `position` in `array`, if we use the `separator` specified.
`String.insertAt(array, newelement, position, separator)`	This function inserts an element (`newelement`) at the specified (0-based) `position` in `array`, if we use the `separator` specified.
`String.removeAt(array, position, separator)`	This function returns the amended `array` with the element specified (0-based) at `position` removed, if we use the `separator` specified.
`String.replaceAt(array, newelement, position, separator)`	This function replaces an element at the specified (0-based) `position` in `array`, if we use the `separator` specified, with the `newelement` supplied.

Table 7-13. Array Manipulation Functions

PRAGMAS

A *pragma* or a *pragma directive* is a command to the compiler that explicitly generates a special behavior from the compiler.

For example,

```
use url UtilityFunctions "utilityfuncs.wmls";
```

tells the compiler to treat any reference in the file to "UtilityFuntions" as a reference to utilityfuncs.wmls This means that simple references can be made throughout a file or application that avoid complex references that invite typographical errors.

If you are going to use any of the pragma directives described in the following sections, you should remember the following:

▼ You must specify any pragmas at the very beginning of the file, before declaring any functions.

▲ All pragma directives start with the keyword use and are followed by their specific attributes.

External Files

It is quite common, when developing an application, to keep functions that you have created in logically discrete units or files. As well as keeping things organized (having all functions that deal with the calculation of interest rates in a file called interest_rates.wmls, for example), you can also break up the process of development across several people, and make each one responsible for the development and maintenance of a specific set of files or code modules.

All you need to do is publish the names of the files and the functions that they contain in a publicly available document, and then you or any other developer can access these functions by telling the compiler that you want to access the external file that contains them.

This is done with the use url pragma, like so:

```
use url UtilityFunctions "utilityfuncs.wmls";
```

This pragma tells the compiler that any reference to UtilityFunctions in the ensuing script means to go bind the reference to the external file utilityfuncs.wmls.

Inside the functions in this script, you can then use just the resource name that you specified. For example:

```
function calcInterest (p_amount, p_rate, p_term) {
if (UtilityFunctions#checkFloat (p_amount) == true) {
    // the amount is a valid floating-point number
}
if (UtilityFunctions#checkFloat (p_ rate) == true) {
    // the rate is a valid floating-point number
}
if (UtilityFunctions#checkNumeric (p_ term) == true) {
    // the loan term is a valid number
}
    // do more code

};
```

The advantages of such an approach should be fairly obvious. All functions only need to be written once and placed in one location on the system, and they can be referenced from there. The alternative is to copy the same functions into every script that needs them, which can cause nightmares. Suppose you found a bug in the function that calculates the interest rates. How many other script files did you copy this function into, and where are they? You have exactly zero minutes to find out and fix them all, because the application went live an hour ago, and one in three prospective customers are getting incorrect quotes. Do you think this never happens? Believe me, it happens all the time.

There are a couple of other points concerning the `use url` pragma that you should be aware of. First, the `use url` pragma has its own namespace for local names, which can add an extra layer of encapsulation to your functions. However, the local names must be unique within any given file.

Second, the URL string can be a fully formed URL, like http://www.test.com/wml/scripts/utilitiyfuncs.wmls, and you should know that no compile-time URL validity checking is done at all. If you have typed the URL incorrectly, you will only find out when you try to run the application.

Access Control

An access-control pragma can be used to protect a file's content by physically restricting the access to a given URL. In this way, you can be confident that any code that is being called that changes or accesses any confidential sections of your database is actually being called from a valid part of your application, and not by a third party.

You can only have one access-control pragma in a script file. Any more than one will generate a compiler error. Also, as with all pragmas, the access-control pragma must be placed at the very top of the script file, before calling any external files.

The syntax of the command is as follows:

```
use access domain "company.com" path "/directorypath";
```

Every time an external function is called, the compiler performs an access-control check to find out if the destination file allows access from the caller. The access-control pragma specifies the `domain` and `path` attributes against which the access-control checks are performed. If a file has a domain or path attribute, the referring file's URL must match the values of the attributes exactly.

To simplify the development of applications that may not know the absolute path to the current file, the `path` attribute accepts relative URLs. For example, if the access-control attributes for a file are these,

```
use access domain "wapforum.org" path "/finance";
```

then the following referring URLs would be allowed to call the external functions specified in this file:

- ▼ http://wapforum.org/finance/money.cgi
- ■ https://www.wapforum.org/finance/markets.cgi
- ▲ http://www.wapforum.org/finance/demos/packages.cgi?x+123&y+456

The following URLs would *not* be allowed to call the external function:

▼ http://www.test.net/finance

▲ http://www.wapforum.org/internal/foo.wml

By default, access control is disabled, so you only need to include this pragma if you want to keep people out of one or more specific script files.

Metadata

These pragmas are used to pass metadata to the compiler. How the metadata is then treated is usually vendor-specific at present. These pragmas specify a file's property name and content. The attribute values passed are string literals.

Meta pragmas can have the following qualifiers:

▼ **Name** The name qualifier specifies metadata used by the origin server. For example, the micro-browser should ignore this:

```
use meta name "Author" "Dale Bulbrook";
```

■ **HTTP equiv** The http equiv qualifier specifies metadata indicating that the property should be interpreted as an HTTP header. For example, this should be converted to a wireless session protocol (WSP) or HTTP response header if the file is compiled before it arrives at the user agent:

```
use meta http equiv "Keywords" "WMLScript, Language, WAP";
```

▲ **User agent** The user agent qualifier specifies metadata intended for the user agents. This must be delivered to the user agent and must not be removed by any network intermediary:

```
use meta user agent "Type" "Test";
```

GENERAL CODING PRINCIPLES

The general principles of writing code in WMLScript are the same as for writing code in any other programming or scripting language. Now that we have covered the basics of the language, it would be irresponsible of me to just leave you to figure out the rest for yourselves—to let you fall into the same traps that every beginning developer has fallen into since computers were invented. There are basic "safety rules" for developing programs.

I am writing here, of course, for the benefit of readers who are new to programming. If you are already a pro, you can just skip ahead to the next chapter.

The following points are the "basic basics"—enough for you to get started with. If they help save you just one day of your life by helping you avoid these situations, I will feel my effort well spent:

- ▼ **Know what you are trying to do before you start to write any code** It is amazing how many people just start to write code to handle a requirement without knowing how they are going to solve the problem. This is also true of experienced programmers who should know better. I am not innocent of this myself, and I wish I had the chance to go back and save all the time I have lost by enthusiastically starting to type and coding myself into a corner. I invariably have to start again from the beginning, after working it all out on paper properly.

- ■ **Build your code incrementally** Don't try to write all of your code at one sitting without testing it as you go along. It is much easier to debug your code one piece at a time than it is to be hit with a dozen bugs at once. Get one function working correctly before you move on. If you write a code block, test and debug it before you write the next one.

- ▲ **Check for *all* possible variations of input** What happens if the user enters a number instead of a character? Or a character instead of a number? Or a punctuation character instead of either a number or a character? One classic error when checking on numbers is to check for "greater than" and then for "less than," and overlook what happens if the number input is "exactly equal to" a number.

The reverse of this is to check for "greater than or equal to" and "less than or equal to," as they obviously will never both be true, and only the first expression that tests to `true` will ever be acted on.

Only by carefully testing all possibilities can we be sure that the application is as correct as possible when it goes live.

CHAPTER 8

Database-Driven WAP

So far, all of the examples and code have been dynamic only in the sense that we have created decks and cards that can respond to input in a usable way. With WMLScript, we can create interactive pages that respond intelligently to user input.

This is all well and good, but the driving force in this day and age is information. With useful information presented to the user in a timely fashion, the user gains that feel-good factor. More importantly, the user can make decisions that can profoundly affect his or her actions based on the information received. This can be as simple as deciding what clothes to wear that day based on the weather forecast, or as complicated as making a "buy" or "sell" decision on a stock holding that could make or save hundreds or even thousands of dollars if the information is truly timely.

Static data hard-coded in cards is fine up to the point that it becomes out of date. Weather information would be a classic example of this. If every time you wanted to change the weather forecast you had to rebuild a card and then upload it to the server, this would get pretty tedious. It would rapidly get much *more* than tedious if you wanted to show the weather for more than one city or region of the country, and keep this updated every hour or two.

And if that looks overwhelming, then what about stocks and shares information where the data is changing by the minute. Keeping such huge amounts of information bang up-to-date with static WML and WMLScript is not desirable or even possible.

The answer to how to do this is contained in database-driven applications. Most of the information in the world is already contained in databases, particularly commercial information that is essential to the running of a company. What we need to do is to tap this information and be able to get it from the database (which is already being kept up-to-date externally) and put it into our cards and decks without our having to intervene in the process.

Nice idea—have the site dynamically updated with current information, while we go and get in a round of golf. Well, dust off the golf clubs, because I am going to show you how to do just that.

Now before we start, I had better tell you that there are several different types of databases, and several different methods of getting data from those databases so that we can use it in the cards, and several different ways of plugging it into the cards themselves. There are Oracle, Access, MS-SQL, MySQL, DBase databases, and many more. Extracting the data and plugging it into the cards can be done with CGI, PHP, ColdFusion, and others.

A few years ago, I had to make a decision. I had to decide whether I was going to continue up the learning curve with Unix, or continue up the learning curve with Microsoft products. One thing was very obvious to me at the time—I did not have the time to keep up with both *and* earn a living.

I decided at that time to go with Microsoft. It is probably much more than a half-truth that the division into the two main "camps" in systems development, Unix and Microsoft, comes about because nobody in the real world has the time to keep up with all of the developments in both, and have a life. Both camps have their upsides and their downsides.

But for the purposes of this book, and because I have much more experience now with Microsoft products than I do with Unix products, I am going to go into detail about how to create a database-driven WAP site using a Microsoft Access database and Active Server Pages.

Although the details will be specific to these products, the principles are universal to all products and platforms.

ACTIVE SERVER PAGES

Active Server Pages (ASP) is a subset of Visual Basic from Microsoft and is a server-side scripting technology, which can be used to create dynamic and interactive web applications. An ASP page is a page that contains server-side scripts processed by the web server before being sent to the user's browser.

A server-side script runs when a browser requests a file with an extension of .asp from the web server. ASP is called by the web server, which processes the requested file from top to bottom and executes any script commands. It then returns the results (whether this is formatted text or database results) to the browser that made the request.

Because the code inside an Active Server Page is executed before the page is sent to the browser, the code itself is never actually seen by the browser.

An additional consequence of this is that the page sent to the browser is standard text and will therefore run normally on all browsers (with the usual caveats on browser compatibility).

Active Server Pages was released in February 1996. Until that time, if you did not have an experienced C programmer on staff or a Unix guru who was familiar with the Perl scripting language, you were extremely limited in even the basic scripting abilities.

What ASP did for Internet web sites is comparable to moving from the horse as a form of transport to motor vehicles. The performance increase is huge.

Let's look at a very simple example.

Here is a Perl script that displays the current time and date:

```
#!/usr/local/bin/perl
print "Content-type: text/plain", "\n\n";
$time=`/usr/bin/date` ;
print "The time is: ", $time;
exit(0);
```

To display the same information using Active Server Pages, you would write

```
<% = time %>
```

The point here is simple. All of the functions from Visual Basic that make ASPs work are already built in, so all you have to do is supply the proper code to accomplish the task at hand.

In addition, Active Server Pages is free and already built into Windows 2000. It is a part of the Internet Information Server (IIS) and can be added with the Windows Add/Remove Programs menu. It also runs just fine with the Personal Web Server (PWS) software available for Windows 95/98/ME.

More importantly, since ASP is native to the Windows NT operating system, it is readily available to all clients with web sites being hosted on the server, and does not require much (if any) intervention from the administrator to access virtually all of its resources.

ASP and WAP

How do we tie up the two technologies of ASP and WAP so that we can use ASP in WAP cards and decks? Fortunately, it is quite simple. Because all of the WML code is hosted on a web server, all we need to do is to put ASP commands into the decks and cards as needed.

By telling the server to then go ahead and process the files as ASP files, the commands are executed and the results sent back to the browser, or in the case of WAP, the micro-browser.

In this way, we can literally add virtually any programming functionality or database interactivity that we like to our decks and cards. This is really exciting stuff, because the only limits we now have on our applications are the physical constraints of the micro-browser itself, as already covered at length.

Here is a very basic example of ASP in action:

```
<% @LANGUAGE="VBSCRIPT" %>
<% Response.ContentType = "text/vnd.wap.WML" %>
<?xml version="1.0"?>
<!DOCTYPE wml PUBLIC "-//WAPFORUM//DTD WML 1.1//EN"
"http://www.wapforum.org/DTD/wml_1.1.xml">

<wml>
  <card id="card1">
    <do type="prev"><noop/></do>
    <do type="Date">
      <go href="#card2" />
    </do>
    <p align="center">
      <b>Ready for some interactive WAP?</b>
    </p>
  </card>
```

```
  <card id="card2">
    <p align="center">
    The date is <%=date()%></p>
  </card>
</wml>
```

As you can see, this looks almost exactly like standard WML. Let's examine the differences (in bold).

First of all, the file *must* be saved with an extension of .asp, so that the server knows to call ASP to process the file before doing anything else with it.

When the file is processed, the first line is

```
<% @LANGUAGE="VBSCRIPT" %>
```

The <% and %> characters delimit script commands in an ASP file and need to be used around the commands; otherwise the ASP commands would be treated like ordinary text. The line above tells ASP to process all commands within the <% %> delimiters in the file as VBScript. If the syntax is incorrect for VBScript, an error will be raised.

NOTE: You could also use JavaScript, but we are going to stick exclusively to VBScript for the purposes of this book. The principles are exactly the same, however, and if you would prefer to use server-side JavaScript, by all means please feel free to do so.

The next line is

```
<% Response.ContentType = "text/vnd.wap.WML" %>
```

The web server will by default return the contents of an ASP file to the browser using the MIME type for HTML. A WAP device or browser will reject this, as it will only accept WML. So we need to override this default and tell the server to return the contents with a MIME type for WML instead. Putting this line at the top of your file will accomplish this.

The rest of the file is completely standard WML, with the exception of the third-to-last line, which is

```
    The date is <%=date()%></p>
```

When ASP processes the file, this script command is executed, and the server date is inserted in the file at this point.

When executed, this deck will look like this:

The first card

The date dynamically inserted

The ASP Object Model

As always, there is more to any technology than first meets the eye, and ASP is no exception. What used to be called the Application Programming Interface (API) to the ASP functionality is now known as the ASP Object Model. This is the set of individual objects that encompass the low-level functions that make ASP the easy-to-use scripting language that it is.

The basic objects that make up ASP that you need to know about at this point are the following: Response, Request, Session, Application, and Server.

Let's quickly go over these one at a time.

The Response Object

The object that returns data to the browser from the server is the Response object. It can do this in immediate mode, as in

```
<% Response.Write "<p>Here is a paragraph</p>" %>
```

which is the usual usage, and which is most often written as

```
<% = "<p>Here is a paragraph</p>" %>
```

with the = sign acting as a shortcut for `Response.Write`.

You can redirect the program flow by using the command

```
<% Response.Redirect "anotherfile.asp" %>
```

and you can also buffer output to build it up before it is sent to the browser by setting

```
<% Response.Buffer = True %>
```

The default behavior for buffering is False, which means that as each line is processed, the result is sent directly to the browser. You can combine these in useful ways, as there may well be times when you do not know what a result of a calculation will be until you are halfway through processing the particular file. You may want to redirect the user to another screen at that point, but if you have already been writing to the browser, you are going to have a mess.

So you could use the Response object to write something like this:

```
<% Response.Buffer = True %>
...
<% ' If there has been an error
Response.Clear    ' Clear the contents of the buffer without sending it.
Response.Write "<p>There has been an Error: Please Try Again</p>"
Response.End     ' Stop processing the page at this point and send it.
%>
```

Incidentally, this is also a very useful debugging trick for examining the contents of variables while you are developing. You can write two lines like the following, and comment them out if not needed by using the single quote (which indicates an in-line comment) at the beginning of the line:

```
<%
Response.Write "<p>Variable is: " & variablename & "</p>"
Response.End
%>
```

The variable is written to the browser, and all processing stops.

The Request Object

The Request object provides us with all of the data about the user's request to the web server or our application. All of this data is read-only, and cannot be modified via the Request object.

The interface gives you access to five collections of variables in the object, and these are (in approximate order of relevance):

- ▼ QueryString
- ■ Form
- ■ ServerVariables
- ■ Cookies
- ▲ ClientCertificate

Of these collections, the first three are the most important, and we will cover the parts of these that we need to know as we come to them.

The Session Object

While Session variables are a great way to store temporary data across different pages in the application, there is a drawback, in that to use them, the browser must support cookies. Cookies are not supported by the WML version 1.1 specifications, although cookies can be supported by the Phone.com gateways. The reason why WAP devices, with very few exceptions, do not support cookies is fairly obvious. They are very costly in terms of memory for a micro-browser, which does not have any real long-term storage, unlike a PC. Storing the cookie on the gateway is a clever solution to this restriction.

While a great deal of personalization can be achieved with the use of cookies, there are other possible ways around this. For example, you could store all of the data that you would normally store in a cookie in a database on the server. When the user logs on, you have all of the personalization data at hand and available for use. It is true that a cookie would enable the device to "log on" without further need to enter a password and so forth, but it can also be argued that this is not very secure for a portable device that can easily be mislaid or stolen.

In any event, if cookies are available, you can store information in Session variables like this:

```
<% Session("User_Security_Level") = 99 %>
```

If this variable had been set after verifying a user's identity, you could then set a security check on each deck of the application that would verify the security level set for that deck, and throw the user out if he or she were not authorized for that deck or module, like this:

```
<% this_deck_security_level = 60 %>
<%
    if Session("User_Security_Level") < this_deck_security_level then
        Response.Redirect "reject.asp"
    end if
%>
<wml>
    ...
</wml>
```

If Session variables are not an option, another alternative would be to check the database for the user's security authorization level at the beginning of every deck. Another more cumbersome method would be to pass any data that you want to make "persistent" from deck to deck with **postfield** elements.

The Application Object

The Application object holds data global to the entire application. The major point to bear in mind here is that while the Session object lasts for the duration of the user's session, the Application object is created the very first time the application is accessed, and persists until the server is rebooted, however long that may be.

If any of the Application variables are modified, they will be changed for all users of the application from that point onward.

For this reason, it is normal to initialize application variables with data that you want to remain constant for the duration of the application.

A good example of this is a database connection string, like so:

```
<% Application("Database_Connection_String") = "dsn=mydatabase;" %>
```

Thereafter, you can connect to the database by referring to the application variable `Application("Database_Connection_String")`. If you want to then change the name of the database (say, when upgrading your application), you only have to change the initialization line of code as given above, and you don't have to touch anything else in the application to have the connection changed.

The Server Object

The Active Server Pages environment can be extended with the use of server components, or COM objects. You will find yourself using the Server object quite extensively, even if it is only to connect to and query databases.

The majority, if not all, of the components you use will be the standard components that come with ASP and the Microsoft environment.

However, you may also write your own customized components with, say, Visual Basic, with your own tailor-made methods and properties, and create instances of these with the Server object just as easily as a built-in component.

The most-used method of the Server object is the CreateObject method, and while we will be seeing a lot more of this later on, it looks like this:

```
<% dim rs, sql
set rs = Server.CreateObject("ADODB.Recordset")
sql = "SELECT * from myTable;"
rs.open sql, Application("Database_Connection_String")
```

At this point, the data returned by the SQL Select statement is ready to be manipulated. This is covered in full in the next section.

ACTIVEX DATA OBJECTS (ADO)

ADO is the object model provided by Microsoft for general data access and manipulation, and is used to make your databases accessible via the Internet. OK, I can hear you now, saying something like, "What, another acronym? What is it this time, and why do I need to know about it when I wanted to find out about WAP, for Pete's sake!" Don't worry, I do understand. If I were in your position as a reader, I would be thinking the same thing.

In fact, one of the frustrating things about being the author of a work such as this is realizing that in order to get to some real meat of a topic we have to go through the gradient scale of background data and theory in order to reach a point where I can say, "See? That's how you become ruler of the universe. Simple, eh?" and have you understand completely how I got to that point.

ADO is one of those things that you have to know enough about to be able to connect to and interrogate a database, and that's all. If you want to know all about ADO, this is not the book for you. But at the end of this section you will know all you need to in order to do some pretty impressive stuff indeed; so bear with me.

What we have seen so far with ASP is pretty powerful, but the ability to connect directly to databases, query data from them, and insert it directly into the card to be displayed to users on their WAP device is a particularly potent one.

For this we need to know something about

- ▼ Making a physical connection to the database itself
- ■ Querying the database for the data we need
- ▲ Accessing and using the data that is returned from the database

This is where ADO comes into the picture.

Just as ASP has its objects, so too does ADO. The ADO objects that we are concerned about are the Connection, Command, and Recordset objects. There are some others, but none that concerns us here.

In fact, ADO is pretty clever, in that if you create a Recordset object, you can simply specify what connection you want to use when you tell the object to perform the query. While this won't work in all cases (Oracle seems to have a limit on the size of the database Select statement with the Recordset object that works just fine if you use a Command object instead), the Recordset object will serve 99 percent of your needs; so I am going to stick with this for the sake of simplicity of demonstration.

Physically Connecting to the Database

First of all, we need to create a Recordset object that will hold the data that we want returned from the database.

This is done with the simple statement

```
<% set RS = Server.CreateObject("ADODB.Recordset") %>
```

We now have a variable, RS, which contains a new instantiation of a Recordset object.

Querying the Database

The next step is to create a statement containing a database query that, when executed, will return the data that will be held in the Recordset object, RS.

Without going into the complexities of SQL, this can be a simple SELECT statement like

```
<% SQLstring = "SELECT * from HOTELS;" %>
```

which simply puts the Select statement as a string into a variable that we have called SQLstring.

The final step is to tell the Recordset object to execute the SQL query and, at the same time, specify how to connect to the database.

This is done by using the Open method of the Recordset object, like so:

```
<% RS.Open SQLString, Application("Database_Connection_String") %>
```

In fact, a "quick and dirty" method of writing the same code would be

```
<%
set RS = Server.CreateObject("ADODB.Recordset")
RS.Open "SELECT * from hotels;", "dsn=mydatabase;"
%>
```

The only problems with writing the code like this are that in any kind of real application you don't want to have to go through all of the code modifying a database name if it changes, and you may well want to build a SELECT statement bit by bit and pass it to the Open method when you have finished. But while it is certainly not good programming practice, the above code will work just fine.

Using the Returned Data

Now that we have the Recordset object created, opened, and (if we got the SELECT statement right) populated with data from the database, we can use the remaining methods of the Recordset object to move through the Recordset, display data, add new records, and maintain existing ones.

If we assume that the HOTELS table contains the fields Hotel_Name and Empty_Rooms, we can list the data very simply by using a loop to iterate through the Recordset like so:

```
<card id="hotels">
<% do while not RS.EOF %>
<p>Hotel: <%=RS("Hotel_Name")%> <br/>
   Rooms: <%=RS("Empty_Rooms")%> </p>
<%
RS.MoveNext
Loop
%>
</card>
```

The above code creates a card called "hotels," and then in a DO WHILE ... LOOP construct, moves through each record in the Recordset object, displaying the hotel name and the number of empty rooms for each record until there is no more data.

This code as written has obvious drawbacks, in that you have to make sure that the data returned in the deck does not exceed the limits of 1K at a time, but the point here is to show you the simplicity of what is needed, not how complicated it can get. You can build up from "simple" to "complicated," but it's much tougher the other way round.

Some of the other methods of the Recordset object are MoveFirst, MoveLast, MovePrevious, Add, Update, and so on.

Tidying Up

When you have finished with the Recordset object, it is good practice to tidy up and close it down, thereby releasing any memory used. This is as simple as writing

```
<%
    RS.Close    ' Close the Recordset object
    set RS = Nothing   ' Release the Recordset objects memory
%>
```

You would normally write this at the end of the deck, but you can put it at any point where you have finished with the Recordset object.

Some Additional Notes about Connections

Because the system overheads of creating a database connection are very high, the designers have cleverly constructed the system so that each connection, once created, is actually left open for a specific period of time after a user has finished with it (this is parameter driven and can be changed by the system administrator) in case either you or another user wants to access the same database with the same details. If such a request is not received within the time period, the connection is closed.

Because the number of simultaneous connections also puts a load on the server, it is good practice to create a Recordset only when you need it, and to close it again as soon as possible after you have finished with it, thus releasing the connection back to the "connection pool" for possible use by another user.

OK, so we have a heap of theory that we have covered here.

Let's go ahead and build something dynamic with it.

CHAPTER 9

A Dynamic WAP Application

Some time ago I built a database application for the clubbing scene, called Worldwide-Dance-Web.com. From this web site, you can select a city or region in the UK, and a night you want to go out, and you get a list of what events are happening at what clubs in that area on that date. Alternatively, you can select a DJ and find out where he or she is playing and on what dates, so that you can dance to your favorite DJ.

What I am going to do is to build a sample WAP application here in front of your very eyes that will provide some of the web site data to a mobile device. This has a major benefit, in that the data is already current and available for the web site; so we are simply enabling a wider section of the public to access this data. For the relatively minor cost of additional development for the mobile market, we are able to reach a much wider audience.

This model will, I am sure, be repeated by most, if not all, of the major database-driven sites for the simple reason that all of the major investment has already been done. The existing site has to be maintained anyway, and the database kept up-to-date; so there is in effect nothing to lose and a lot to gain.

For the sake of simplicity, I am only going to build a single module, which will be the selection of the club you want to go to by selecting the DJ that you want to spend the night dancing to. The remaining options, like the selection of the venue by Area and/or Date, the News screens, the latest Charts listings, the e-commerce module for buying DanceWeb T-shirts and music files, can all be added separately at a later time.

Let's get this working so we can convince the owners of the site to go ahead and commission the remaining parts of the application.

WORLDWIDE-DANCE-WEB FOR WAP

Like any other application, this has to be planned out in advance, and then built. We will cover in detail the processes of working out the data flow so that we can see what will be needed from the data processing, building the database required to handle the data needed, and then writing the code itself so as to handle the user interactions.

Data Flow

The first step is to plan the flow of the screens for the application, and in that way we can see what will be needed from the back end of the application.

Starting from the home page of default.wml, we attempt to log in to the application. If we are successful, we go through to the DJ Menu screen, menu.asp, and either select a DJ (in which case we go to the results screen, results.asp) or we can opt to change our preferred DJs. If we decide to change our preferences, we are given a selection of the preference "slots" to change in prefs.asp. If we choose a slot, we are able to add a DJ to that slot, and our changes are written to the database in wrprefs.asp. The data flow is shown in the following illustration.

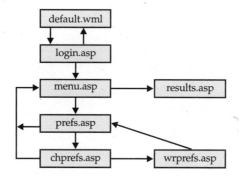

Building the Database

The database tables are quite straightforward, and for the purposes of this chapter I have created vastly stripped down versions of the actual tables used for the live web site. The first table is the Users table, and the user names and passwords are kept in here for login authentication.

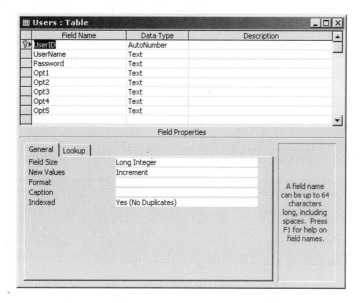

Now we need to have a table called DJ which contains the DJ names for us to list to the User when he decides to modify his preferences.

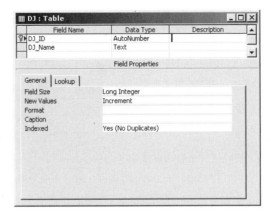

We now create the Venue table, which holds the basic data on the club itself, like address, phone number and so forth.

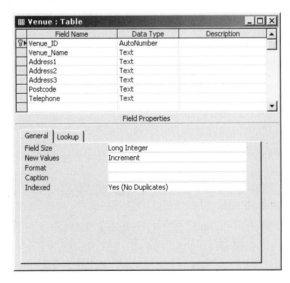

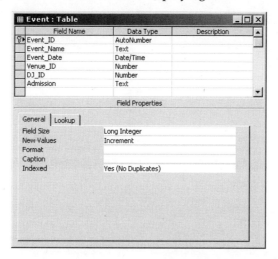

Finally, we need to have a table with the information on each of the "events", or club nights that are held, with the details of date, DJ playing, etc.

Creating a DSN

Last of all, we will need to create a DSN (data source name) entry for our database. This creates a system entry for the database, which allows the system to physically locate the actual database via a single name that you specify.

Open the Control Panel and double-click the ODBC Data Sources 32 bit (or similarly named) icon. This will display the ODBC Data Source Administrator. Click the System DSN tab, and click the Add button. Choose the Microsoft Access (*.mdb) driver and click Finish (see the following illustration).

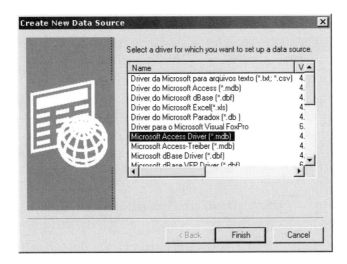

The ODBC Microsoft Access Setup dialog should appear (see the next illustration). Type in the data source name for this database (in this case, **wapdb**). Use the Select button to identify where you have saved the database on your machine. If you now click the OK buttons on each of the two dialog boxes to close them, you have finished creating the DSN.

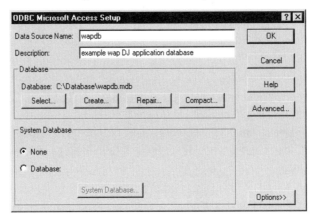

Writing the Code

Our first card will be straight WML. You can type this into Notepad. I actually used Microsoft Visual Interdev to create the ASP pages, and it works just fine. If you also install M3Gate, you can view the pages and debug as you go. Using an external editor like this also gets you away from the defaults that, say, the Nokia editor imposes on you, and gives you more freedom.

Default.wml

As the name indicates, this is the default or Home page for the application. This is the first page that the user will see, if the server has been set up to look for a file with this name automatically.

```
<?xml version="1.0"?>
<!DOCTYPE wml PUBLIC "-//WAPFORUM//DTD WML 1.1//EN"
"http://www.wapforum.org/DTD/wml_1.1.xml">

<!-- user welcome and login -->
<wml>
<head>
    <meta http-equiv="Cache_Control" content="max-age=0"/>
</head>
<card>
    <do type="accept" label="Login">
        <go href="login.asp?User=$(User)&amp;Pwrd=$(Pwrd)"/>
    </do>
    <p align="center">
    <b>DJ Finder</b>
    </p><p>
    <br/>User ID:
    <input name="User" maxlength="10" type="text" emptyok="false"/><br/>
    </p><p>
    <br/>Password:
    <input name="Pwrd" maxlength="10" type="text" emptyok="false"/><br/>
    </p>
</card>
</wml>
```

For this scenario, we have assumed that the user is already registered. We can add another option to this page that will take the user to a sign-up or registration page, but we can do this later. We could also default users to a "guest" account, but then they would not be able to access the personalization options from the database.

In the meantime, the previous is a fairly standard login card that collects a user name and password to send back to the server. When users click "Login", they are taken to the file login.asp. The only element that we have not covered yet is the line

```
<meta http-equiv="Cache_Control" content="max-age=0"/>
```

This tag makes sure that the card is not read from the cache after the user has finished with it, but instead that the deck is re-read from the server. Using this tag ensures that there are no problems with old content if the user has to come back and enter his or her ID again. It also ensures that anyone picking up a lost phone will not automatically have access to the password by looking through the cache.

NOTE: I have made the password data entry field a text type, not a password type as you would on a PC.

You should be able to access this card by using the URL http://localhost/default.wml in your browser, or whichever site you have created to hold the project.

The next pages start to use some serious interactive code.

Login.asp

This ASP script is used to process the user ID and password that have been entered in default.wml. It checks that the user name and password are on the database and sends the appropriate response back to the client device.

```
<% @LANGUAGE="VBSCRIPT" %>
<% Option Explicit %>
<% ' Handle login verification
Response.Expires = 0

Dim rs, Username, Passw, Query, Success

Username = Request("User")
Passw = Lcase(Request("Pwrd"))

Set rs = Server.CreateObject("ADODB.Recordset")
Query = "select UserID, UserName from users " & _
        "where username = '"&Username&"' and password = '"&Passw&"'"

rs.Open Query, "dsn=wapdb;"
%>
<%Response.Buffer = True%>
<%Response.ContentType = "text/vnd.wap.wml"%>
<?xml version="1.0"?>
<!DOCTYPE wml PUBLIC "-//WAPFORUM//DTD WML 1.1//EN"
"http://www.wapforum.org/DTD/wml_1.1.xml">

<wml>
<head>
    <meta http-equiv="Cache-Control" content="max-age=0"/>
</head>
<card>

<% if rs.EOF then%>
    <do type="accept">
```

```
        <go href="default.wml" />
    </do>
    <do type="prev" label="Back">
        <prev/>
    </do>
    <p align="center">
    <br/>Invalid Login!
    <br/>Please try again.
    </p>

<%Else%>
<%
Session("UserID") = rs("UserID")
Session("User") = rs("UserName")
%>

    <do type="accept">
        <go href="menu.asp"/>
    </do>

    <p>Welcome <%=rs("UserName")%>!</p>
    <p>You are now logged in.</p>

<%End If%>

</card>
</wml>

<%
rs.Close
set rs = nothing
%>
```

If everything behaved itself, and cookies are supported on your WAP gateway, you should now be able to log on with one of the user names and passwords that you have entered into the Users table, and get the message "Welcome "xxx"! You are now logged in." The "xxx" is the user's name as retrieved from the database.

Let's break this code down and see how it works. The line

```
Dim rs, Username, Passw, Query, Success
```

declares five variables that will be used later in the file. It is mandatory for all variables to be declared before being used if the "option explicit" statement is used. Although it is not necessary, it is always a good idea to use this. If you accidentally misname a variable later in the file, you will get an error message when you try to run the code, and it will tell you what you have done wrong. Without the "option explicit" statement, you would simply not get the results that you originally expected and have to go through a possibly painful debugging session to find out where you went wrong.

The next two lines show the use of the Request collection of the ASP Object Model covered earlier:

```
Username = Request("User")
Passw = Lcase(Request("Pwrd"))
```

Because the parameters were passed in the URL string, they can be referred to by the target file with "Request" as shown in the previous code. In this case, they are being stored in two of the variables that were declared earlier. Strictly speaking, this is not necessary, and you could refer to them directly in the rest of this file by simply using `Request("User")` where you needed it. However, there is an overhead to using object collections, whereas use of straightforward variables does not carry this penalty. Their use once or twice in the file is no big deal, but in the case of large files with possible referencing of the same variables many times, it is much cleaner to use a local variable.

The next lines are the ones that set up and open the query to the database table, just as previously covered:

```
Set rs = Server.CreateObject("ADODB.Recordset")
Query = "select UserID, UserName from users " & _
        "where username = '"&Username&"' and password = '"&Passw&"'"

rs.Open Query, "dsn=wapdb;"
```

The database query in this instance is getting a list of all records in the database where the Username field and the Password fields are identical to those entered by the user. I have split the lines for the sake of legibility in print, using the line continuation character ("_"), but you can put it all on one line if you prefer. The result is stored in the **rs** Recordset object.

The next two lines have already been covered in an earlier section:

```
<%Response.Buffer = True%>
<%Response.ContentType = "text/vnd.wap.wml"%>
```

The "Response.Buffer" is set to True so that no content is sent to the WAP device until the page has been completely processed, and the "ContentType" is set to "wml" so that the WAP device will accept the content.

If there are records in the Recordset, the record pointer is placed at the first available record, and the End of File (EOF) property is set to False. If there are no records in the Recordset, both the Beginning of File (BOF) and End of File (EOF) properties are set to True. This means that you can always check the End of File property to see if there is more, or any, data.

The next ASP line demonstrates this:

```
<% if rs.EOF then%>
```

If the user name and password are found, there will be one record on file. If not, the End of File property will be True, and we can test it as shown here. If it is True, the user name and password were not found, and we can handle the user accordingly. In this case, we tell the user that it was an invalid login, and to please try again. If this is the case, then we also only allow the user to go back and try again.

The next part should also be pretty clear:

```
<%Else%>
<%
Session("UserID") = rs("UserID")
Session("User") = rs("UserName")
%>
```

Now that we have handled the case where the user is not on the database, the next step is to handle the successfully registered user. The first thing that we have done here is to set up two Session variables to store the user ID and the user name. In this way, we can access this data from anywhere else in the application without having to go back to the database.

The final step after building the card that allows the user into the rest of the application is to close down the Recordset object gracefully. You can get away with just closing the deck and leaving the objects to handle themselves, but this is definitely "quick and dirty," and not recommended coding practice. You should always have your application in a known state. Leaving it to the operating system of any server or device is always risky, as there are no guarantees.

We have now built the access to the site, so we have to do something with the users now that they are in.

Menu.asp

This ASP file, or deck, will construct a menu for the users based on their own preferences, and also add menu options that will allow them to add to or change existing preferences.

```
<% @LANGUAGE="VBSCRIPT" %>
<% Option Explicit %>
<% ' Present the User with a menu
Response.Expires=0

Dim rs, Query, t_Name, t_User

t_User = Session("UserID")
t_Name = Session("User")
```

```
Set rs = Server.CreateObject("ADODB.Recordset")
Query = "SELECT opt1, opt2, opt3, opt4, opt5 FROM users " & _
        "WHERE userid = "&t_User&""
rs.Open Query, "dsn=wapdb;"
%>

<%Response.Buffer = True%>
<%Response.ContentType = "text/vnd.wap.wml"%>
<?xml version="1.0"?>
<!DOCTYPE wml PUBLIC "-//WAPFORUM//DTD WML 1.1//EN"
http://www.wapforum.org/DTD/wml_1.1.xml">

<wml>
<head>
    <meta http-equiv="Cache-Control" content="max-age=0"/>
</head>
<card>
<do type="prev" label="Back">
<prev/>
</do>
<p align="center">
<b><%=t_Name%>'s Menu:</b></p>

<p>
<%If rs("opt1")<> "" then%>
<a href="results.asp?action=<%=rs("opt1")%>"><%=rs("opt1")%></a>
<br/>
<%End If%>

<%If rs("opt2")<> "" then%>
<a href="results.asp?action=<%=rs("opt2")%>"><%=rs("opt2")%></a>
<br/>
<%End If%>

<%If rs("opt3")<> "" then%>
<a href="results.asp?action=<%=rs("opt3")%>"><%=rs("opt3")%></a>
<br/>
<%End If%>

<%If rs("opt4")<> "" then%>
<a href="results.asp?action=<%=rs("opt4")%>"><%=rs("opt4")%></a>
<br/>
<%End If%>
```

```
<%If rs("opt5")<> "" then%>
<a href="results.asp?action=<%=rs("opt5")%>"><%=rs("opt5")%></a>
<br/>
<%End If%>

<a href="prefs.asp">Preferences</a>
<br/>

<a href="default.wml">Home</a>
<br/>
</p>
</card>
</wml>
<%
rs.Close
set rs = nothing
%>
```

We have already covered most of the points in this file, with just a couple of exceptions. The first one is setting local variables from the Session variables we created in login.asp. The two lines are

```
t_User = Session("UserID")
t_Name = Session("User")
```

As mentioned before, it is better to create and access local variables wherever possible. These lines just take the contents of the Session variables and put them into local variables.

To embed the ASP code into the body of the WML document, we can use the shortcut form of "Request.Write," as in the next line:

```
<b><%=t_Name%>'s Menu:</b></p>
```

Here, the local variable t_Name is written directly into the file at this point. As this is the variable that holds the user's name, what will actually appear on the WAP device (if the user's name is Dale) will be "Dale's Menu:"

Now we have to build the menu itself, and we do this by checking on the contents of each option field one at a time. If the field is not empty—that is, there is an option held in the field—then we want it displayed. But we don't just want it displayed as text; we want to have it displayed as a clickable link that will take the user to another deck as a result of clicking that option. The lines are

```
<%If rs("opt1")<> "" then%>
<a href="results.asp?action=<%=rs("opt1")%>"><%=rs("opt1")%></a>
<br/>
<%End If%>
```

As you can see, the first line checks to see if the "opt1" field is empty. If it is not, we construct a link element that passes a variable called `action` to the deck results.asp with the value of the field "opt1."

If the field *is* empty, we write nothing at all. In this way, we can build a menu that contains only the user's options, with no dummy statements or spaces on the display. There are five of these IF statements, or one for each option.

If we had designed the database differently—for example, with a table that links to the user ID and allows any number of options instead of restricting it to five—we could have just built the menu with a simple loop through the second table, but we would also have had to create a second Recordset. In the interests of simplicity I have done it this way. We will see a more complex type of interaction later, in results.asp.

If we run the application with what we have done so far, you will see that we have a menu appear with just two items on it—Preferences and Home. So far, so good. Now let's allow the users to add their preferences.

Prefs.asp

The prefs.asp file generates a list of the current menu selections as links to the file chprefs.asp, which is the file that does the actual work of changing the preferences themselves.

```
<% @LANGUAGE="VBSCRIPT" %>
<% option Explicit %>
<% ' Capture the users preferences
Response.Expires=0

Dim rs, Query, t_Name, t_User

t_User = Session("UserID")
t_Name = Session("User")

Set rs = Server.CreateObject("ADODB.Recordset")
Query = "SELECT opt1, opt2, opt3, opt4, opt5 FROM users " & _
        "WHERE userid = "&t_User&""
rs.Open Query, "dsn=wapdb;"
%>

<%Response.Buffer = True%>
<%Response.ContentType = "text/vnd.wap.wml"%>
<?xml version="1.0"?>

<!DOCTYPE wml PUBLIC "-//WAPFORUM//DTD WML 1.1//EN"
          "http://www.wapforum.org/DTD/wml_1.1.xml">

<wml>
<head>
```

```
      <meta http-equiv="Cache-Control" content="max-age=0"/>
</head>
<card>
<do type="prev" label="Back">
<prev/>
</do>
<p align="center">
<b><%=t_Name%>'s DJ's:</b></p>

<p>
<a href="chprefs.asp?action=opt1">1 = 
<%If rs("opt1")<>"" then%>
<%=rs("opt1")%></a>
<%Else%>
Empty </a>
<%End If%>
<br/>

<a href="chprefs.asp?action=opt2">2 = 
<%If rs("opt2")<>"" then%>
<%=rs("opt2")%></a>
<%Else%>
Empty </a>
<%End If%>
<br/>

<a href="chprefs.asp?action=opt3">3 = 
<%If rs("opt3")<>"" then%>
<%=rs("opt3")%></a>
<%Else%>
Empty </a>
<%End If%>
<br/>

<a href="chprefs.asp?action=opt4">4 = 
<%If rs("opt4")<>"" then%>
<%=rs("opt4")%></a>
<%Else%>
Empty </a>
<%End If%>
<br/>

<a href="chprefs.asp?action=opt5">5 = 
<%If rs("opt5")<>"" then%>
```

```
<%=rs("opt5")%></a>
<%Else%>
Empty </a>
<%End If%>
<br/>

<a href="menu.asp">Menu</a>
<br/>
<a href="default.wml">Home</a>
<br/>

</p>
</card>
</wml>
<%
rs.Close
set rs = nothing
%>
```

The code should be starting to look quite familiar now. The major difference between the chprefs.asp file and prefs.asp is that we are displaying every option on the screen, even if they are empty.

```
<a href="chprefs.asp?action=opt1">1 = 
<%If rs("opt1")<>"" then%>
<%=rs("opt1")%></a>
<%Else%>
Empty </a>
<%End If%>
```

The idea behind this is that if users click on any of the menu options, they are sent to the chprefs.asp file, along with the option number that they have chosen as a parameter, so that chprefs.asp knows which option to change.

We are displaying the option as "Empty" so that the user can select the option to give it a value. The only time we would hit a problem with this is if there happens to be a DJ named "Empty."

The next step is to create the chprefs.asp script.

Chprefs.asp

The chprefs.asp file is where we can change the users preferences for a favorite DJ or set of DJs, so that they can be recalled for this user without the user having to do anything else the next time they access this application.

```
<% @LANGUAGE="VBSCRIPT" %>
<% Option Explicit %>
```

```
<% ' Change User DJ Choices
Response.Expires=0

Dim rs, Query, t_Opt, t_Choice

t_Opt = Request("action")

Set rs = Server.CreateObject("ADODB.Recordset")
Query = "SELECT DJ_Name FROM DJ ORDER BY DJ_Name;"
rs.Open Query, "dsn=wapdb;"
%>

<%Response.Buffer = True%>
<%Response.ContentType = "text/vnd.wap.wml"%>
<?xml version="1.0"?>
<!DOCTYPE wml PUBLIC "-//WAPFORUM//DTD WML 1.1//EN"
"http://www.wapforum.org/DTD/wml_1.1.xml">

<wml>
<head>
    <meta http-equiv="Cache-Control" content="max-age=0"/>
</head>
<card>
<do type="prev" label="Back">
<prev/>
</do>
<p align="center">
<b><%=t_Opt%>'s DJ Choice:</b></p>
<p>

<%Do While Not rs.EOF%>

<%t_Choice=rs("DJ_Name")%>

<a href="wrprefs.asp?action=<%=t_Choice%>&amp;opt=<%=t_Opt%>">
<%If rs("DJ_Name ")<>"" then%>
<%=rs("DJ_Name ")%></a>
<%Else%>
Empty </a>
<%End If%>
<br/>

<%rs.MoveNext
Loop%>
```

```
<a href="menu.asp">Menu</a>
<br/>
<a href="default.wml">Home</a>
<br/>

</p>
</card>
</wml>
<%
rs.Close
set rs = nothing
%>
```

The first difference in this code is the line

```
<%Do While Not rs.EOF%>
```

This loop will go through the file and create a list of DJs that can be selected to fill the selected option. The list is simply a list of links in this application. The loop is completed with the lines

```
<%rs.MoveNext
Loop%>
```

The final step in the process of changing preferences is to write the selected option into the user's selected option number in the user record.

Wrprefs.asp

This file is different, in that it does not send anything to the WAP device at all. It is "pure" code that performs a specific task, and when completed it redirects the program flow back to prefs.asp so that the user can make another choice for another option if wanted. The code is as follows:

```
<% @LANGUAGE="VBSCRIPT" %>
<% Option Explicit %>
<% ' Write the User choices back to the file
Response.Expires=0

Dim rs, Query, t_Opt, t_Choice, t_User
Const adOpenStatic = 3
Const adLockOptimistic = 3

t_Opt = Request("opt")
t_Choice = Request("dj")
t_User = Session("UserID")
```

```
if t_User = "Make Empty" then
    t_User = ""
end if

Set rs = Server.CreateObject("ADODB.Recordset")
Query = "Select * from Users where userid="&t_user&";"
rs.Open Query, "dsn=wapdb;", adOpenStatic, adLockOptimistic

rs(t_opt) = t_Choice

rs.Update

rs.Close
Set rs = nothing
Response.redirect ("prefs.asp")
%>
```

Note that there are two variables declared as constants at the top of the file: `adOpenStatic` and `adLockOptimistic`. These are used in the Recordset Open command later in the file. The point to note here is that these constants are actually declared in an "include" file, adovbs.inc, which you could insert at the top of the file with the following command if you wanted to.

```
<!-- #include file="adovbs.inc" -->
```

I personally don't like to do this for two reasons. First, there are a *lot* of constants declared in the file adovbs.inc, and it seems like a waste of processing time to add them all. Second, you have to remember the names of the constants and type them each time you refer to them.

I have included the names of the constants here so you can see that in order to write to a Recordset object to update the file, you need to declare the type of Recordset object you want (Static), and you also have to declare that you want record locking enabled in order to be able to write to the file. You do not need to specify these if you are just reading from the file, as record locking is not an issue then.

You could in fact just write

```
rs.Open Query, "dsn=wapdb;", 3, 3
```

and it would work just fine. However, this negates the entire reason for using these constants in the first place, which is to allow Microsoft to change the numbers at any time. As long as programmers always use the constants file and the variable names, no changes to the code should ever have to be made.

Realistically though, if Microsoft were to change the value of the constants in the constants file after so long with no changes, an awful lot of code would break all across the planet as many programmers write code using the values of the constants directly, just as I have done.

Getting back to the file, the user variables are picked up and stored to local variables. There is nothing new in that, but there is an IF statement:

```
if t_User = "Make Empty" then
    t_User = ""
end if
```

What this does is allow the user to select an option called "Make Empty," (which we have not added to the database, but it is just another entry), and if the user does select it, the option is set to a null string, effectively deleting it from the User menu.

We then open the file as described previously, which allows us to update the Recordset for this user's record.

The next line replaces the data in the file for the selected option:

```
rs(t_opt) = t_Choice
```

This line just says to put the string in "t_Choice" into the field in the Recordset collection that is held in the variable t_opt (which of course is going to be one of "opt1," "opt2," "opt3," "opt4," or "opt5").

And the final line that we are interested in here is

```
rs.Update
```

which tells the Recordset object to actually do the update to the record itself.

NOTE: Using the Recordset object is not necessarily the optimum way to do updates, but it *is* workable, and I have used it for consistency with the other files in the application.

The only thing left to do now is to have something useful happen when the user clicks on the DJ name from his or her personal DJ menu. We have already set this up to call a file called results.asp, and it is described next.

Results.asp

This is the file that does all the "real" work, and displays the data for the DJ that was requested by the user after it has been read from the database.

```
<% @LANGUAGE="VBSCRIPT" %>
<% Option Explicit %>
<% ' display the next event that the DJ is playing
Response.Expires=0

Dim rs, Query, t_Name, t_User, t_DJ
```

```
t_User = Session("UserID")
t_Name = Session("User")
t_DJ = Request("dj")

set rs = server.CreateObject("ADODB.Recordset")

Query = "SELECT DJ.DJ_Name, Event.Event_Name, Event.Event_Date, " & _
"Event.Admission, Venue.Venue_Name, Venue.Address1, " & _
"Venue.Address2, Venue.Address3, Venue.Postcode, Venue.Telephone " & _
"FROM Venue INNER JOIN (DJ INNER JOIN Event ON " & _
"DJ.DJ_ID = Event.DJ_ID) ON Venue.Venue_ID = Event.Venue_ID " & _
"WHERE (DJ.DJ_Name)='" & t_DJ & "' " & _
"ORDER BY Event.Event_Date;"

rs.Open Query, "dsn=wapdb;"
%>

<%Response.Buffer = True%>
<%Response.ContentType = "text/vnd.wap.wml"%>
<?xml version="1.0"?>
<!DOCTYPE wml PUBLIC "-//WAPFORUM//DTD WML 1.1//EN"
          "http://www.wapforum.org/DTD/wml_1.1.xml">

<wml>
<head>
    <meta http-equiv="Cache-Control" content="max-age=0"/>
</head>
<card>
<do type="prev" label="Back">
<prev/>
</do>
<p align="center">
<small><u>Next Event for <%=t_DJ%></u></small></p>

<p>
<small>

<%if rs.EOF then %>
Sorry, we don't have the details of this DJ's next event yet.
Please try another.<br/>

<%Else%>

Event: <%=rs("Event_Name")%>
```

```
<br/>
Date: <%=rs("Event_Date")%>
<br/>

Venue: <%=rs("Venue_Name")%>
<br/>
Address: <%=rs("Address1")%>
<br/>

<%if rs("Address2") > "" then%>
    <%=rs("Address2")%>
    <br/>
<%end if%>
<%if rs("Address3") > "" then%>
    <%=rs("Address3")%>
    <br/>
<%end if%>

<%=rs("Postcode")%>
<br/>
Phone: <%=rs("Telephone")%>
<br/>
Admission: <%=rs("Admission")%>
<br/>

<%End If%>
</small>

<a href="prefs.asp">Preferences</a>
<br/>
<a href="menu.asp">Menu</a>
<br/>
</p>
</card>
</wml>
<%
rs.close
set rs = nothing
%>
```

The query that is set up here pulls data from the Venue and Event tables, based on the DJ ID that is looked up from the DJ table. It might look complex, but the easiest way to create a query is simply to build it in Access, and then cut and paste it into your ASP page or deck. This works very well, as you can modify it and test it to your heart's content within Access without having to retest your whole application just to get the query statement right.

You might notice in this file that I do not loop through any records in the Recordset. This is quite intentional. If there are several events in the table for this DJ, the file size (not to mention user scrolling) could very quickly become excessive. Instead, I am sorting the records into date sequence, and then only ever referring to the first record, which is the next event for this DJ. (You could, and should, put in additional code here to double-check that an old record hasn't been deleted, and is displayed in error.)

If the Recordset is empty, we know that there are no events currently on file for this DJ, and a corresponding error message is displayed. Otherwise, the fields are displayed in a logical sequence down the screen. Later on, we will briefly revisit the results page again, to enable the user to dial the telephone number displayed on the screen by simply clicking on it.

SUMMARY

As you can probably tell, the application begun in this chapter is a long way from being fully complete. However, I hope that it has enabled you to see just how an application can be built in ASP.

While I have not touched on any other scripting languages, as mentioned at the beginning, the principles are the same regardless of the development platform. If you are familiar with Oracle, or MySQL, or PHP, or even Perl, you will immediately recognize how you can write the code presented here with tools that you are more comfortable with.

CHAPTER 10

Converting Existing Web Sites

Now we know how to do everything we need to build a site in WML. That's fine if we are going to be designing and building a site from scratch. However, it is a much more likely scenario that, initially at least, we will be converting an existing site to WML so that the owners of the site can reach a wider market. In this case, there are advantages and drawbacks to doing a site conversion.

The major advantage is that the basic logic of the site—the navigation structure, the content, the brand name, and so forth—has already been set up.

The major disadvantage is that the site will have been designed and built with, and around, HTML design constraints. If the site uses frames, for example (and many do), we instantly have a problem. There is no equivalent in WML to the `<frameset>` tag, so we are immediately in for some redesign and creative thinking for that specific site.

WHY CONVERT AN EXISTING HTML WEB SITE TO WAP?

There are now millions of existing web sites that have already been written in HTML. All of this content is factually only available to users with access to a web browser. What are the benefits of converting an existing web site to WAP? The instant benefit is that the already large number of users now potentially able to view your content will increase massively. There are various predictions floating around, with various degrees of believability, but it is certain that in the long run there will be far more people with mobile WAP-enabled devices than people who own computers.

In the UK, anyone going into most major chain stores and supermarkets is met by huge displays of cheap mobile phones, the majority of which are WAP enabled. This means that somebody who does not own a computer can walk into a supermarket, buy a phone, do a search via whatever portal the phone has been set up with, and find your site. As mobile devices become more sophisticated, and as the prices of the combination phone/PDAs come down and allow instant voice communications, more people will be able to locate and access your site.

What Should You Convert?

The ideal sites for conversion are those that have a simple navigation structure and information that can be summarized in a few sentences. If the site is already large, with lots of deep links and massive amounts of text, it may be necessary to design and build a WML version from scratch. This obviously will require more time, effort, and consequently, expense than converting from an existing HTML site. If your site has lots of graphics, or if audio is a vital component of the site, it will be very difficult, if not impossible, to convert in any kind of satisfactory way.

Dynamic pages—pages that change often—can be handled by keeping the scripting (ASP or otherwise) sections in the code. If the site is HTML and changes often, you will have to set extra time aside to maintain the WML version of the site as well as the HTML version.

If there is a lot of text content, but it is held in a database, one possible solution (which I demonstrate later in this chapter) is to add another field or fields to the database specifically for the WAP content only. When updating or adding records to the database, it is a relatively trivial task to then abstract the larger text contents to these additional fields. In this way, both HTML and WML sites can still be updated from the same database with no further maintenance necessary once the conversion has been done.

Finally, you should take into account the content of the existing site and consider whether it is relevant for a mobile market in the first place. While a "Property Search" for a four-bedroom house might be fine for somebody on the daily commute by train, giving people access to government white papers via a micro-browser might be considered of benefit only to masochists and opticians.

Methods of Conversion

There are two methods of conversion that we will touch lightly on here, as they place the resulting WML code outside of your control, and hence outside of the scope of this book. They are worth mentioning because they provide an "instant fix." The results, however, are comparable to walking into a travel agent and saying, "Give me a vacation," then accepting whatever you get.

You *might* get a month in Hawaii. Unfortunately, you probably won't.

These two methods of conversion are fully automated converters and configurable converters.

Fully Automated Converters

Fully automated converters simply pick up the original source code and convert it to WML based on a simple set of rules, then squirt it out directly to the micro-browser. You never get to see or touch the converted code. This is better than nothing, but the results can leave a lot to be desired.

Because the conversion is done at the WAP gateway, you have absolutely no control over the final appearance of the display at the micro-browser.

In addition, at the moment, the only gateway to provide this service is the Phone.com gateway. If users are accessing the site from another gateway, they get nothing—not really a consistent or desirable state of affairs if your company's public image is important.

Configurable Converters

Very simplistically, a configurable converter allows you to put instructions into the HTML as a set of tags, usually within comments, that tell the converter how to handle a particular section of code. In this way you can be very specific with relevant sections. For example, a common feature of many order forms on a web site is to allow the user to select the country of origin from a pick list. This would be quite prohibitive in terms of deck size for a micro-browser.

One alternative to this would be to provide an abbreviated list for the converter to display that only contains the most likely countries, not all of them.

Configurable converters can also do things like convert the HTML to formats other than WML, which makes them more useful than the fully automated converters because they can be updated to use any kind of format that may come out in the future.

Finally, note that both of these methods of conversion are provided only by third-party products, which can be expensive and restrictive in terms of options.

So what other option is left?

Do-It-Yourself—The Hands-On Approach

Until the research into artificial intelligence produces a piece of software that can work out just what the designers intended, and write WML and WMLScript that will do exactly that, we are left with the "other alternative," or Plan B—that is, to automate what can be automated and tweak manually what cannot be automated, or what the automation has gotten wrong.

None of the automatic tools will handle JavaScript, or VBS, or ASP, or Perl, or in fact anything additional to the basic HTML tags. You will still have to manually convert or create any scripts that you need as WMLScript files, convert any images you want to WBMP files, and generally tidy up.

The only simple conversion tool freely available at this time is the ArgoGroup WAP Tool, available from http://www.argogroup.com.

This tool will do a lot of the work of manually converting a web page for you, but it will only handle HTML files at this time. You can remove the ASP script tags (<% %>) and rename the file to .htm, and it will convert the raw HTML for you, but trying to handle any files with scripts in this way is not really fast, smooth, or automatic. I do use it for straight HTML files though, as it converts the files according to the rules laid out in this chapter. This saves a lot of time. The WML files can then be tweaked to handle any excess text, images, and so on. (Incidentally, the ArgoGroup tool also has a bonus, in that it automatically converts the HTML to XHTML as well, and you can save the resultant XHTML file as well as the WML file. You will appreciate this feature if you have ever had the tedious and repetitive task of converting an HTML page to XHTML.)

There is another alternative, and that is to use the data in this chapter to write your own conversion program. If you are already a programmer, you only need to write routines to parse the HTML or ASP file and search for and replace any tags as given in Table 10-1. This has some major advantages. First, you can customize the conversion to your own exact specifications, and add company-specific data or "meta" tags, as you like. Second, you can precisely control the conversion process and automate it as you see fit. If you want to run your conversion program against the HTML/script files every Monday at 9:00 A.M., you can do so and be able to predict the exact output WML files with no reliance on any third party.

In other words, the whole conversion process becomes yours to own and control, and in today's uncertain world, that is no small thing.

Table 10-1 contains the basic rules that you need to follow in order to convert from HTML to WML, based on what we have learned so far. The same rules will apply to automatic or manual conversion.

NOTE: Don't forget, as WML is a strict XML specification, the usual restrictions apply: all WML tags must be in lowercase, all attributes must be in quotes, and all tags must be explicitly closed.

HTML Tag	WML Tag
<HTML>	<wml>
<HEAD>	<head>
<META>	<meta>
<TITLE>	<title>
<BODY>	<card><p></p></card>
<P>	Only the text inside the <P> tag needs to be transferred, as the <p></p> tags for the card have already been created for the <body> (see above).
<A>	<a>
<UL></UL>	Unordered list. There is no WML equivalent. You can create a bulleted point (with *, for example), followed by the text of the element.
<OL></OL>	Ordered list. There is no WML equivalent. You can create a bulleted point (with *, for example), followed by the text of the element.
<LI></LI>	List item. There is no WML equivalent. You can create a bulleted point (with *, for example), followed by the text of the element.
<B></B>	<strong></strong>
<EM></EM>	<em></em>
<STRONG></STRONG>	<strong></strong>
<U></U>	<u></u>

Table 10-1. HTML to WML Conversion Rules

HTML Tag	WML Tag
 	 A special note here: it seems that while (with a space) will work in all tested browsers, (with no space) will work in some but not others. This applies to <hr /> also.
<HR>	<hr /> (See note on .)
<TABLE><TR><TD></TD></TR><TABLE>	There is no guaranteed WML equivalent. If you convert the content of the table data <td> as text, you can format the lines with spaces or punctuation characters as needed.
<FORM> (If there is a "submit" input type)	If the form method is "post," you should use <do type='accept' label='submit'> <go href =action method="post"> <postfield> … < nth postfield> </go> </do> If the form method is "get," you should use <do> <go href=action method="post"> </do> The "action" attribute should take all the necessary values according to the inputs.
<INPUT>	<input>
<INPUT type="radio">	Radio button types should be converted to <select>/<option> tags with single selection, like so: <select> iname=name ivalue=1 (for Radio) <option value="value" > </option> </select>

Table 10-1. HTML to WML Conversion Rules *(continued)*

HTML Tag	WML Tag
<INPUT type=" checkbox">	Checkbox button types should be converted to <select>/<option> tags with multiple selection, like this: form <select> iname=name ivalue=2 (for Checkbox) <option value="value" > </option> </select>
<SELECT> Attributes: name value size multiple	<select> Attributes: iname=name ivalue=value title =name If you use the 'multiple' attribute, then you must write multiple=true Or multiple = false
<OPTION> Attribute Value	<option> Attribute Value=Value
<TEXTAREA>Text</TEXTAREA>	<input type="text" value="Text"/>
<IMG>	<img alt="alt text" /> Note that the "alt" attribute is mandatory in WML.
<INPUT type="button">	<do> type='accept' label=value <go> href =value </go> </do>
<FONT>	There is no WML equivalent.

Table 10-1. HTML to WML Conversion Rules *(continued)*

HTML Tag	WML Tag
<SCRIPT>	There is no direct WML equivalent. If you are writing your own converter, you could write all scripts into a separate file with an extension of .wmls for later modification.
<% %> (ASP) or other script tags	If the file is still to be executed as an ASP or similar file, these tags can be left alone and included as is in the output file.
<DIV>	There is no WML equivalent.
<CENTER>	<p align="center">
Any text not in a paragraph tag...	<p>Any text not in a paragraph tag</p>
<APPLET>	Cannot convert

Table 10-1. HTML to WML Conversion Rules *(continued)*

The tags listed in Table 10-1 will pretty much cover most HTML or script files. I will probably be spending some time writing such a conversion program once I have finished this current project, and I will make it as generic as possible for future use.

A DEMONSTRATION HTML CONVERSION

Let's take a look at an example of a small conversion of a couple of ASP files that output details of holiday destinations. The actual text content is quite large, as the company obviously wants to sell the customer on the joys or beauties of a particular resort or hotel. What I have done is to add another text field to the database that will only be used in the WML implementation, which is a greatly reduced version, but still gives vacation seekers the incentive either to phone the holiday company right away or visit the web site when they can get to a PC for full details.

Figure 10-1 is a snapshot of part of the database for one property with a ShortDescription field added. And the following is a listing of the file we are going to convert—properties.asp. Don't be concerned if you can't follow the HTML. Although

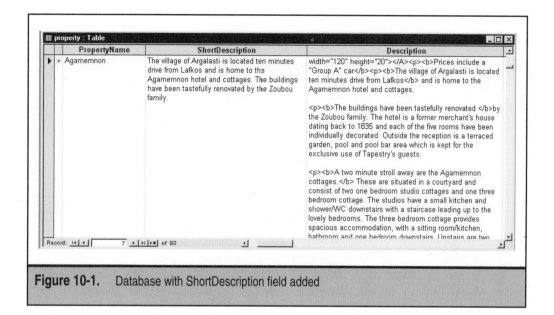

Figure 10-1. Database with ShortDescription field added

this is from a live system that my company put together, we are using it here (with permission) as a demonstration only.

```
<% @language=VBScript %>
<%
Session("Property") = Request("property")

set RS = Server.CreateObject("ADODB.Recordset")
RS.ActiveConnection = "dsn=tapestry;"

RS.Source = "SELECT property.*, OfferProperties.OfferID " & _
"FROM property LEFT JOIN OfferProperties ON " & _
"property.PropertyID = OfferProperties.PropertyID " & _
"WHERE (((property.PropertyName)='" & Session("Property") & "'));"
RS.CursorType = 0
RS.CursorLocation = 2
RS.LockType = 3
RS.Open

%>
```

```
<HTML>
<HEAD>
<META http-equiv="pragma" content="no-cache">
<LINK rel="stylesheet" type="text/css"
      href="<%=Session("Country")%>.css">
<SCRIPT language="JavaScript" type="text/javascript">
<!--
<!-- .
function EFix(){
for (a in document.links)
   document.links[a].onfocus = document.links[a].blur;}
if (document.all){document.onmousedown = EFix;}
// -->

function MM_swapImgRestore() { //v3.0
var i,x,a=document.MM_sr;
for(i=0;a&&i<a.length&&(x=a[i])&&x.oSrc;i++) x.src=x.oSrc;
}

function MM_findObj(n, d) { //v3.0
  var p,i,x;  if(!d) d=document;
  if((p=n.indexOf("?"))>0&&parent.frames.length) {
    d=parent.frames[n.substring(p+1)].document; n=n.substring(0,p);}
  if(!(x=d[n])&&d.all) x=d.all[n];
     for (i=0;!x&&i<d.forms.length;i++) x=d.forms[i][n];
     for(i=0;!x&&d.layers&&i<d.layers.length;i++)
         x=MM_findObj(n,d.layers[i].document);
  return x;
}

function MM_swapImage() { //v3.0
  var i,j=0,x,a=MM_swapImage.arguments;
  document.MM_sr=new Array; for(i=0;i<(a.length-2);i+=3)
   if ((x=MM_findObj(a[i]))!=null){document.MM_sr[j++]=x;
   if(!x.oSrc) x.oSrc=x.src; x.src=a[i+2];}
}

function MM_openBrWindow(theURL,winName,features) { //v2.0
    window.open(theURL,winName,features);
}
//-->
</SCRIPT>
<TITLE>Properties</TITLE>
<BASE target="_self">
</HEAD>

<BODY >
<H1><IMG src="images/spacer.gif" width="128" height="14"><BR>
   Properties</H1>
```

```
<% if not RS.EOF then%>
<TABLE width="100%" border="0" cellspacing="1" cellpadding="1">
  <TR>
    <TD valign="bottom" align="left" colspan="5">
      <H2><%=RS("PropertyName")%>    
<%if ucase(RS("Elite")) = "YES" then %>
<A href="#" onMouseOut="MM_swapImgRestore()"
onMouseOver="MM_swapImage('Elite','','elite1.gif',1)">
<IMG name="Elite" border="0" src="elite2.gif" width="71" height="20"
onMouseDown="MM_openBrWindow('elite.htm','Elite','width=500,height=350')">
</A>
<%end if%>
    
<%if RS("OfferID") > "" then %>
<A href="#" onMouseOut="MM_swapImgRestore()"
onMouseOver="MM_swapImage('Spech','','speshoff1.gif',1)">
<IMG name="Spech" border="0" src="speshoff.gif" width="71" height="20"
onMouseDown="MM_openBrWindow('lateoffers.htm','LateOffers')"></A>
<%end if%>
</H2>
    </TD>
  </TR>
  <TR>
<TD colspan="2"><A href="resorts.asp" onMouseOut="MM_swapImgRestore()"
onMouseOver="MM_swapImage('GotoResorts','','goresort2.gif',1)">
<IMG name="GotoResorts" border="0" src=" goresort1.gif" width="120"
height="20"></A>
    </TD>
    <TD> <A href="regionmap.asp?region=<%=Session("region")%>"
onMouseOut="MM_swapImgRestore()"
onMouseOver="MM_swapImage('GotoRegion','','goregion2.gif',1)">
<IMG name="GotoRegion" border="0" src="goregion.gif" width="120"
height="20"></A>

    </TD>
<TD><A href="resort_selector.asp" onMouseOut="MM_swapImgRestore()"
 onMouseOver="MM_swapImage('GotoCountry','','gocountry2.gif',1)">
<IMG name="GotoCountry" border="0" src="gocountry1.gif"></A>
    </TD>
    <TD colspan="2"> </TD>
  </TR>
  <TR>
    <TD valign="top" align="left" colspan="2">
 <% if Session("country") = "Turkey" then %>
     <A href="viewflightsturkey.asp" onMouseOut="MM_swapImgRestore()"
onMouseOver="MM_swapImage('ViewFlight','','viewflights2.gif',1)">
<IMG name="ViewFlight" border="0" src="viewflights1.gif"></A>
      <% end if%>
<% if Session("region") = "Cephalonia" then %>
```

```
<A href="viewflightscephalonia.asp" onMouseOut="MM_swapImgRestore()"
onMouseOver="MM_swapImage('ViewFlight','','viewflights2.gif',1)">
<IMG name="ViewFlight" border="0" src="viewflights1.gif"></A>
      <% end if%>
<% if Session("region") = "Pelion" then %>
<A href="viewflightspelion.asp" onMouseOut="MM_swapImgRestore()"
onMouseOver="MM_swapImage('ViewFlight','','viewflights2.gif',1)">
<IMG name="ViewFlight" border="0" src="viewflights1.gif"></A>
      <% end if%>
</TD>
    <TD align="left" valign="top">
<A href="viewprices.asp" onMouseOut="MM_swapImgRestore()"
onMouseOver="MM_swapImage('ViewPrices','','viewprices2.gif',1)">
<IMG name="ViewPrices" border="0" src="viewprices1.gif"></A></TD>
<TD align="left" valign="top">
<FORM action="addtolist.asp">
<INPUT type="hidden" name="propertyid" value="<%=RS("PropertyID")%>">
<INPUT type="image" src="addprefs.gif" name="submit" width="120" height="20">
</FORM>
    </TD>
    <TD align="left" valign="top"> </TD>
  </TR>
  <TR>
    <TD valign="top" align="left" colspan="5" height="10">
      <HR noshade size="1">
    </TD>
  </TR>
</TABLE>

<TABLE width="100%">
  <TR>
    <TD align="left" valign="top" colspan="2">
      <TABLE width="20%" border="0" align="right">
        <TR align="right">
         <TD><%if RS("Image2") > "" then %>
             <IMG src="<%=RS("Image2")%>" border="1">
             <%end if%>
         </TD>
        </TR>
        <TR align="right">
          <TD><%if RS("Image3") > "" then %>
             <IMG src="<%=RS("Image3")%>" border="1">
             <%end if%>
          </TD>
        </TR>
        <TR align="right">
          <TD><%if RS("Image4") > "" then %>
             <IMG src="<%=RS("Image4")%>" border="1">
             <%end if%>
```

```
            </TD>
          </TR>
          <TR align="right">
            <TD><%if RS("Image5") > "" then %>
                <IMG src="<%=RS("Image5")%>" border="1">
                <%end if%>
            </TD>
          </TR>
        </TABLE>
        <%=RS("Description")%></TD>
    </TR>

  <%End if%>
</TABLE>

</BODY>
</HTML>
```

Now if we apply all of the rules outlined above, and leave the ASP scripts intact, we get this version—properties.wml:

```
<% @language=VBScript %>
<%
Session("Property") = Request("property")

set RS = Server.CreateObject("ADODB.Recordset")
RS.ActiveConnection = "dsn=tapestry;"

RS.Source = "SELECT property.*, OfferProperties.OfferID " & _
"FROM property LEFT JOIN OfferProperties ON " & _
  "property.PropertyID = OfferProperties.PropertyID " & _
"WHERE (((property.PropertyName)='" & Session("Property") & "'));"
RS.CursorType = 0
RS.CursorLocation = 2
RS.LockType = 3
RS.Open

%>
<%Response.ContentType = "text/vnd.wap.wml"%>
<?xml version="1.0"?>
<!DOCTYPE wml PUBLIC "-//WAPFORUM//DTD WML 1.1//EN"
          "http://www.wapforum.org/DTD/wml_1.1.xml">
<wml>
<card title="Properties">
<p>
<onevent type="onenterforward">
<refresh>
<setvar name="propertyid" value="RS('PropertyID')"/>
</refresh>
```

```
</onevent>
<onevent type="onenterbackward">
<refresh>
<setvar name="propertyid" value="RS('PropertyID')"/>
</refresh>
</onevent>
</p>
<p>
<big><%=RS("PropertyName")%>    
<%if ucase(RS("Elite")) = "YES" then %>
<A href="#" onMouseOut="MM_swapImgRestore()"
onMouseOver="MM_swapImage('Elite','','elite1.gif',1)">
<IMG name="Elite" border="0" src="elite2.gif" width="71" height="20"
onMouseDown="MM_openBrWindow('elite.htm','Elite','width=500,height=350')">
</A>
<%end if%>
    
<%if RS("OfferID") > "" then %>
<A href="#" onMouseOut="MM_swapImgRestore()"
onMouseOver="MM_swapImage('Spech','','speshoff1.gif',1)">
<IMG name="Spech" border="0" src="speshoff.gif" width="71" height="20"
onMouseDown="MM_openBrWindow('lateoffers.htm','LateOffers')">
</A>
<%end if%>
</big>

<big><b><u>
<img alt="Image without an 'alt' attribute" src="images/spacer.gif"/>
<br/>
Properties</u></b></big>
</p>
<p>
<%if not RS.EOF then%>
</p>
<p/>
<p>
<a href="resorts.asp">
<img alt="Image without an 'alt' attribute" src="goresort1.gif"/>
</a><br/>
<a href="regionmap.asp?region=<%=Session("region")%>>
<img alt="Image without an 'alt' attribute" src="goregion.gif"/>
</a><br/>
<a href="resort_selector.asp">
<img alt="Image without an 'alt' attribute" src="gocountry1.gif"/>
</a>
</p>
<p>
<a href="viewflightsturkey.asp">
<img alt="Image without an 'alt' attribute" src="viewflights1.gif"/>
```

```
</a>
<a href="viewflightscephalonia.asp">
<img alt="Image without an 'alt' attribute" src="viewflights1.gif"/>
</a>
<a href="viewflightspelion.asp">
<img alt="Image without an 'alt' attribute" src="viewflights1.gif"/>
</a> <br/>
<a href="viewprices.asp">
<img alt="Image without an 'alt' attribute" src="viewprices1.gif"/>
</a><br/>
<do label="submit" type="accept">
<go href="addtolist.asp" method="get">
<postfield name="propertyid" value="$propertyid"/>
</go>
</do>
</p>
<p align="center">
-----
</p>
<p>
<%=RS("Description")%>
</p>
</card>

</wml>
```

This now gives a "rough cut" or intermediate stage of code. After final tweaking and removal of redundant image code, we wind up with the final version:

```
<% @language=VBScript %>
<%
Session("Property") = "Agamemnon"

set RS = Server.CreateObject("ADODB.Recordset")
RS.ActiveConnection = "dsn=tapestry;"

RS.Source = "SELECT property.*, OfferProperties.OfferID " & _
"FROM property LEFT JOIN OfferProperties ON " & _
    "property.PropertyID = OfferProperties.PropertyID " & _
"WHERE (((property.PropertyName)='" & Session("Property") & "'));"
RS.CursorType = 0
RS.CursorLocation = 2
RS.LockType = 3
RS.Open

%>
<%Response.ContentType = "text/vnd.wap.wml"%>
<?xml version="1.0"?>
<!DOCTYPE wml PUBLIC "-//WAPFORUM//DTD WML 1.1//EN"
```

```
            "http://www.wapforum.org/DTD/wml_1.1.xml">
<wml>
<card title="Properties">
<do type="prev">
<prev />
</do>
<%if not RS.EOF then%>
<p>
<do label="submit" type="accept">
<go href="addtolist.asp" method="get">
<postfield name="propertyid" value="<%=RS("PropertyID")%>"/>
</go>
</do>
</p>
<p>
<%=RS("ShortDescription")%>
</p>
<%end if %>
</card>
</wml>
```

A page from the web site is shown in Figure 10-2, and the output from the WML version looks like Figure 10-3. Although it is a long way from the original web page, it should still be enough to intrigue the user looking for a holiday in Greece.

SUMMARY

As you can see, it might be tedious to convert from HTML to WML, but it is not particularly difficult in theory. You should now be able to tackle the majority of HTML sites without a great deal of difficulty. As mentioned earlier, the hardest part will be solving navigation problems, chiefly for sites based on framesets.

Chapter 10: Converting Existing Web Sites 211

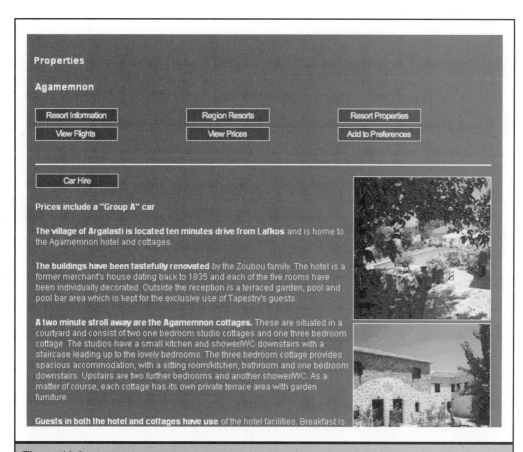

Figure 10-2. Web site screen capture

Figure 10-3. WML version

CHAPTER 11

M-Commerce and Security

Without some kind of secure method of exchanging money via computers, and proving beyond a shadow of a doubt that the participants in the transaction are who they say they are, it would be impossible to buy or sell anything at all over the Internet. The entire network would be useful for basic communication and entertainment only. Since our entire livelihood as developers depends in a very real sense on the security of the transactions that we generate, it is a good idea to know a little about the principles involved.

TYPES OF SECURITY AND WHY IT IS NECESSARY

Let's look at an overview of security and the different levels of security required for the different kinds of things that you might normally do every day, and for some kinds of things that are not of the everyday variety. For example, if you were asking a question about the weather, you wouldn't particularly care at all who overheard you. If you were having an extramarital affair, you would be quite concerned about who had access to your communication lines, but only really concerned about a very limited number of people.

If you were developing a company strategy that took sales away from a competitor, you would be very concerned about who could see that strategy, and you would be prepared to take serious measures to keep it out of the public domain. After all, if the competitor were to find out your strategy from any source, they could take steps to outflank you and even increase their sales at your expense.

If you were attempting to hide certain facts from the population of a country in order to manipulate them into making certain decisions for whatever purpose (this is called politics), or if you were attempting to manipulate a foreign power into taking a certain course of action for trade or military purposes (which is politics again, but known as foreign policy), you would be extremely concerned indeed about who was able to access the relevant data. This is why every government on the planet has the equivalent of the UK's Official Secrets Act.

In a time of war, the last thing you would want at all would be the enemy gaining access to your overall strategies and current troop deployments, and you would be quite justified in becoming extremely paranoid about security.

And unfortunately in the current society, criminality has reached a point where ordinary citizens have to worry about whether their credit cards and personal details have been obtained by someone who will use them without their consent.

The Japanese have a saying that "business is war," and they operate accordingly. The individual or group that has the most data about their competitor, or opponent, or enemy, can, if they are reasonably intelligent, win the day. Therefore, it is realistic to say that some data has an actual value and needs to be protected just as you would protect money or other valuables.

What Is an Acceptable Level of Security?

It was Benjamin Franklin who said, "Two can keep a secret—if one of them is dead." There is no such thing as perfect security, and anything that has been encrypted can be decrypted, given enough time and resources. So security, or encryption, is defined in terms of strength. At one end of the spectrum, you can have no security or "weak" security, and at the other end you have "very strong" security.

An acceptable security level is almost always a compromise between the usability and the strength of the encryption method. It is not possible to define certain standards for an "acceptable" level of security, because the level needed is always dependent on the data being transferred. We have to assume that the transmitted information has some value, but the owner of the information needs to decide which security level is acceptable to preserve the confidentiality of the information.

A company's strategic plan may require strong encryption methods. On the other hand, a short message to friends asking them to join you for lunch may not require any encryption at all. Sending a list of stock prices would require little or no encryption, but sending a buy or sell instruction might require medium to strong encryption. You could say that you might as well strongly encrypt all messages, and then you wouldn't have to worry about it. But encryption is more than just selecting an algorithm. Different algorithms have different resource requirements in order to work properly. The general rule of thumb is that the stronger the selected algorithm, the more computing resources it requires.

From the Wireless Transport Layer Security (WTLS) point of view, the acceptable security level always requires a trade-off with the usage of limited resources. There is no point in using a large percentage of the limited computing resources for encryption and decryption, or in creating unnecessary excess traffic on the narrow bandwidth. However, the WTLS has to ensure a certain security level in order to be used for commercial purposes.

How Secure Is WAP?

The measure of security in WAP depends on what you are trying to do and what you consider secure. For general day-to-day use, WAP is as secure as the digital mobile phone standard that your phone works on, which is basically very secure against eavesdropping. So if you were interested in buying, for example, theater tickets over a WAP device, it is fairly secure. The signal from your WAP device to the WAP gateway using the WTLS, and from the WAP gateway onward to the Internet is as secure as a normal e-commerce web site.

However, there is a fraction of a second at the WAP gateway when the information has been converted from WTLS to plain text, and before it has been encrypted to TLS, that it can in theory be captured by someone with direct access to the WAP gateway. From a bank's point of view, that is not secure enough for bank use. But for day-to-day use, giving your credit card details over a WAP device should be as secure as using your credit card to buy something direct from your PC. In this case, it really all boils down to how much you trust the owners of the gateway you are using.

Some of the big banks in the UK are already offering mobile banking services that allow customers to keep track of their finances via their WAP device, offering details of recent account transactions, balances, and checkbook ordering, with more complex services being added as they become available. This is being done by adding extra layers of security with Digital Certificates specific to the bank itself. To understand what this means, we need to take a brief side trip into the history books.

A BRIEF HISTORY OF ENCRYPTION

The discovery of public key cryptography in the late 1970s was possibly the most significant event in the involvement of large companies and corporations in the Internet, and hence the rapid expansion of the Internet itself. Until these techniques were invented, the use of cryptography was limited to certain groups (like military, governmental, and financial institutions), which had the money and motivation to invest heavily in protecting their information, as well as the personnel and procedures necessary to transport keys securely from one place to another.

Public key cryptography changed all that. It made it possible for any computer user to send a message to any other computer user without fear of eavesdropping. Secure communication via computer is available to everyone.

Cryptography

Cryptography is the science of concealing information. The word itself is defined as "the art of writing or deciphering messages in code." The derivation is "crypto-," meaning secret or hidden, and "-graphy," meaning writing or recording.

For most of history, there was only one practical form of cryptography, known as *symmetric ciphers*. A simple example of a symmetric cipher is the cipher used by Julius Caesar, which simply moves every letter in the alphabet up by a number of places. For example, we could encrypt

This is the message

by moving every letter up three places in the alphabet, to get

Wklv lv wkh phvvdjh

In this case, the text "Wklv lv wkh phvvdjh" is called the *ciphertext*, and the number 3 (the number of places each letter was moved up) is the *key* to the cipher. The person who receives the ciphertext has to know the key to decrypt the message—in this case, by moving each letter back three places in the alphabet.

This particular cipher is obviously very weak because there are only 26 possible keys, but it illustrates what a symmetric cipher looks like. It's called symmetric because anyone who knows the encryption key (move forward three places) also knows the decryption key (move back three places), and vice versa. All symmetric ciphers have this property, though the ones in use today have many more possible keys than the Caesar cipher does.

Symmetric ciphers are very useful, but they have one major drawback. The drawback is that both the person sending the message and the person receiving it have to know what the key is. And that means, at some point, the key has to be transported from one of them to the other in a secure way.

During World War II, spies and their bases had to have shared codebooks. Later in the war, scientists developed specialized hardware that could be operated inside tamperproof boxes and shipped from place to place by trusted couriers. Once both parties knew the key, they could communicate securely whenever they wanted. But establishing the key in the first place was an expensive and risky business.

Symmetric cryptography made secure communications possible for people who really wanted it—governments, the military, large financial institutions—but it was never going to be easy enough for members of the public to use.

Public Key Cryptography

In the early 1970s, three Stanford academics, Whitfield Diffie, Martin Hellman, and Ralph Merkle, had an idea. What if knowing the encryption key *didn't* mean that you automatically knew the decryption key? What if it was so hard to find the decryption key from the encryption key that you could publish your encryption key openly? Then, anyone who wanted to send you a message could simply look up your published encryption key, encrypt the message with that, and send it to you. Only you would know what the corresponding decryption key was; only you could decrypt the message. No one else could crack it, even if they knew the publicly published encryption key. This was the basic idea of public key cryptography.

Having the idea was one thing, but finding a way to make it work was another. Diffie, Hellman, and Merkle tried to come up with a workable method. They managed to find a method, known as the Diffie-Hellman key agreement, for two people to agree on a shared, secret symmetric key using only public messages. It was nice, but it wasn't public key encryption. The discovery of a true public key encryption system was left to Ron Rivest, Adi Shamir, and Leonard Adelman, who were working at MIT. Their system, named the RSA cryptosystem after them, was based on the difficulty of finding the divisors of large numbers (a process known as factoring).

Not only does the public key method let others send you encrypted messages, but it has another nice property: if you encrypt a message with your private key, only your public key can decrypt it properly. This means you can prove that a message came from you by encrypting it with your private key. If it decrypts properly with your public key, then you must have sent it. If it doesn't decrypt properly, it must have come from someone else. This process, known as *digital signing*, provides you with a way to establish your identity without needing a signature on a piece of paper.

Diffie and Hellman had originally envisaged public keys being held in an open directory, where they could be looked up as necessary when people wanted to send secure messages or check signatures. Nowadays it's more common for public keys to be obtained in the form of *digital certificates*, or digitally signed documents that guarantee that a given public key belongs to a given person or company. The result in both cases is the same. A trusted third party creates a link between the public key, which could have been created by anyone, and the identity of its owner, in the same way that a government department (a trusted third party) issues a passport that binds a signature and a photograph to the identity of a person.

One of the major barriers to the widespread usage of public key cryptography was speed of processing. Public key cryptography relies on "hard" mathematical problems—problems that, even with the best current techniques, take a long time to solve. These types of mathematical problems also involve using types of data that current computer architectures can't handle efficiently, for example, numbers that are 129 digits (500 bits) in length or even longer. When public key cryptography was first invented, it was simply too slow to be widely deployed. Fortunately, Moore's Law has essentially solved the problem.

NOTE: In 1965, Gordon Moore was preparing a speech and made a memorable observation. When he started to graph data about the growth in memory chip performance, he realized there was a striking trend. Each new chip contained roughly twice as much capacity as its predecessor, and each chip was released within 18 to 24 months of the previous chip. If this trend continued, he reasoned, computing power would rise exponentially over relatively brief periods of time. Moore's observation, now known as Moore's Law, described a trend that has continued and is still remarkably accurate. It is the basis for many planners' performance forecasts.

On standard PCs, the performance hit from a single public key operation is almost imperceptible. Even on smaller devices, like PDAs, public key operations can easily be performed. And in general, standards have been designed so that public key cryptography is kept to a minimum—it's usually used just to establish your identity initially and agree to a symmetric algorithm and key. These are then used to do most of the heavy work, since symmetric key operations require a lot less processor power than public key operations.

Another major barrier to the use of public keys was the export regulations of several governments. Because of cryptography's military origins, and because cryptography can be used by unscrupulous people to prevent law enforcement agencies from intercepting criminal communications, its sale overseas was restricted. For a long time, the U.S. government (among others) wouldn't permit the export of cryptographic software unless the keys in it were kept artificially short. This was bad for cryptography in two ways. First, it meant that the software from the United States (the world's largest software producer) that was sold in the rest of the world didn't provide adequate protection against eavesdropping. Second, it made it very hard to trust any cryptographic software, from anywhere, sold anywhere.

Following a long campaign by the software industry, the export regulations in the United States have been gradually relaxed. Software producers are now able to export strong cryptography relatively unimpeded.

OK, so that's the history. How does this tie up with WTLS?

WIRELESS TRANSPORT LAYER SECURITY

The security layer in the WAP architecture needs to enable services to be extended over mobile networks while preserving the integrity of the user data. One of the most important requirements is to support low data transfer rates. For example, the Short Message Service (SMS) as a bearer can be as slow as 100 bits per second; so the amount of overhead has to be kept as small as possible because of this low bandwidth.

Other issues include slow interactions, limited processing power, and memory capacity. The round-trip times can be long because of latency, and the connection shouldn't be closed because of that. For instance, the time between the request and response using the SMS bearer can be as long as ten seconds.

Any cryptography algorithms that are used must be "light" enough so that the mobile devices can run them. For example, the amount of available RAM in the mobile devices should be taken into account, as well as the raw computing power of the CPU available.

Shown in Figure 11-1 is the list of protocols that together make up the Wireless Application Protocol, so that you can see where the WTLS fits. The WTLS layer operates above the transport protocol layer, and it provides the upper-level layers of the WAP with a secure transport service interface. The interface preserves the transport interface below it, and it also provides methods to manage secure connections.

The WTLS provides end-to-end security between the Wireless Application Protocol end points. In fact, the end points are the mobile device and the WAP gateway. When the WAP gateway makes the request to the origin server, it will use the Secure Sockets Layer (SSL) below HTTP to make that transaction secure. This is the point where the data is decrypted and again encrypted at the WAP gateway, as mentioned earlier.

The complete secure connection between the client and the service can be achieved in two different ways. The safest way is for the service providers to place a WAP gateway in their own network. Then the whole connection between the client and the service can be trusted because the decryption will not take place until the transmission has reached the service provider's own network, and not in the mobile operator's network.

The service and content providers can also trust the mobile operator's gateway and use virtual private networks to connect their servers to the WAP gateway. In this event, they do not have the ability to manage and control the parameters used by the WTLS at the WAP gateway.

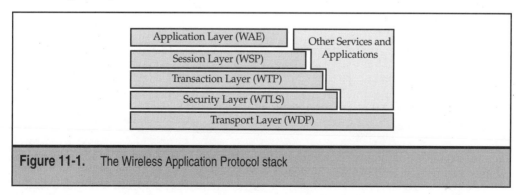

Figure 11-1. The Wireless Application Protocol stack

The Handshake

The negotiating parties can decide the security features they want to utilize during the connection. According to the security requirements, the applications can enable and disable different features of the WTLS. For instance, "privacy" may be left out if the network already provides this service.

All of the security-related parameters are agreed upon during the handshake. These parameters include attributes such as protocol versions to be used, the cryptographic algorithms to be used, and which authentication and public key techniques are to be used to generate a shared "secret." (The "secret" is the *key* for the symmetric encryption described previously.) Figure 11-2 shows the full handshaking sequence.

The handshake starts with the client sending a "Client Hello" message to the server. The server responds to the message with a "Server Hello" message. In the two hello messages, communicating parties agree on the session capabilities. For example, the client announces the supported encryption algorithms and the trusted certificates known by the client. The server responds by determining the session properties to be used during the session. If the client does not suggest any, the server decides on them.

After the client has sent the "Client Hello" message, it starts receiving messages until the "Server Hello Done" message is received. The server sends a "Server Certificate" message if authentication is required on behalf of the server. Additionally, the server may require the client to authenticate itself. The "Server Key Exchange" is used to provide the client with the public key that can be used to conduct or exchange the premaster secret value.

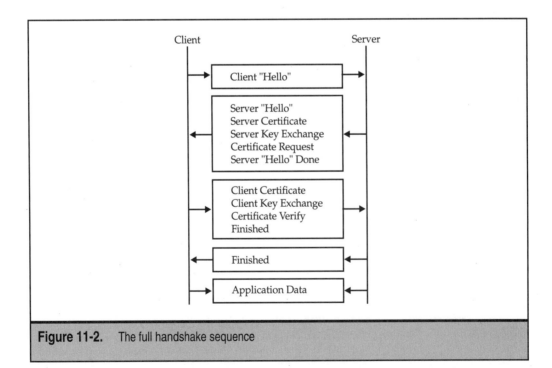

Figure 11-2. The full handshake sequence

CHAPTER 12

Push Technology and Telematics

First of all, I need to explain what push technology and Telematics are before we get into any details of how to make them work for us. I will also explain the basics of the push framework and show how push and Telematics together will make a major difference to consumers businesses in the future.

PUSH TECHNOLOGY

The idea behind *push technology* is very simple. Users can subscribe to desired content, such as the news, and content providers can then send new or current and up-to-date information without a specific request from users. The normal Internet and WAP model is that information is provided only when requested by the client, either by clicking on a hyperlink or by entering a URL into the browser. Simplistically, it looks like Figure 12-1. The push model, on the other hand, states that information can be originated by the server and sent to the client device *without* a request by the client, as shown in Figure 12-2.

The marketing departments of companies across the world were rubbing their hands in glee at the advent of push technology. Instead of trying to get the customer to come to them, they could "push" their message right in front of the customer. Consequently, when it was first released, push technology seemed to be part of every content provider's long-term strategic plan. Like many other ideas that are great in theory, however, push technology on the Internet never really took off with users.

Some of the problems that faced push technology on the Internet were the existence of similar, easier options (like opt-in e-mail lists), the fact that business users were often not in front of their computers when fresh information was pushed, and that home users paying for their phone connection by the minute didn't always appreciate having one or two hundred kilobytes worth of data shoved at them when they only logged on to get their e-mail.

Also, the fact that the data wasn't actually requested in "real time," plus the extremely pertinent fact that a majority of the channels available were not related to "on-the-job" production, meant that system administrators often disabled push on their networks, either on their own initiative or on instructions from senior management.

Although it never really took off on the Web, there are several reasons why push technology is better suited to WAP. In fact, push makes much more sense for mobile devices than it does for desktop computers. A mobile device is normally switched on and at hand all the time. There is no real use of e-mail that would compete with push. And the WAP 1.2 specification includes a complete infrastructure for delivering push content to mobile devices, which provides a standard that can be achieved by all content providers.

The Push Framework

The push framework included in WAP 1.2 is quite complex and probably more than you need to know from a book for beginners, but I am including a brief overview so that you can at least discuss the subject intelligently. If you want to get more technical, you can look up the excellent documentation on the topic at Phone.com and the WAPForum. The framework itself looks like Figure 12-3.

The push framework defines Push Initiators (PI), which would be implemented by the push content providers, and Push Proxy Gateways (PPG), which are responsible for delivering the pushed content to the WAP client. There is also a Push Access Protocol (PAP), which is how the Push Initiator communicates to the Push Proxy Gateway, and a push Over-The-Air (OTA) protocol for communicating between Push Proxy Gateways and WAP devices.

Since the Push Proxy Gateways need to implement the entire WAP protocol, the Push Access Protocol, and the Over-The-Air protocol, it makes sense to combine the push gateway with the WAP gateway; but this is not mandatory. The Push Access Protocol is what is used by the Push Initiator to communicate with these push gateways over the Internet.

The push Over-The-Air protocol is used for communications between the Push Proxy Gateway and WAP devices. And as we are dealing with the WAP/WML realm, you should note that the Over-The-Air protocol rides on top of the WAP "WSP" layer.

Push Messages

The Push Access Protocol defines six types of messages, or operations:

- ▼ Push submission
- ■ Result notification
- ■ Push cancellation
- ■ Status query
- ■ Client-capability query
- ▲ Bad-message response

All of the message types, except the bad-message response, are made up of a pair of messages, one from Push Initiator to Push Proxy Gateway, and the response in the reverse direction.

The push submission operation includes a push message that is sent from a Push Initiator to the Push Proxy Gateway, and a push response back from the Push Proxy Gateway to the Push Initiator. The push message contains the actual information that is being sent to the WAP device. The push response contains a response result code that tells the Push Initiator if the push message was successfully delivered to the Push Proxy Gateway. The push message also contains the information necessary for the Push Proxy Gateway to identify the WAP device that is being targeted over the Over-The-Air protocol.

This identification is done by specifying a device address. The device address can either be a traditional IP number or a Mobile Station International Subscriber Directory Number (MSISDN). This is the international standard for specifying mobile telephone numbers.

The result-notification operation includes a result-notification message sent from the Push Proxy Gateway to the Push Initiator, and a result-notification response sent back from the Push Initiator to the Push Proxy Gateway. The result-notification message will give the Push Initiator details about the success or failure of the delivery of the push submission to the actual device. The result-notification response will tell the Push Proxy Gateway whether the Push Initiator successfully processed this result-notification message. This operation is important, as it is how the Push Initiator is notified of a condition that caused a delivery failure. The three most obvious cases are when users are out of range of the mobile network, have their mobile device switched off, or are already busy with another call.

The push cancellation operation is used for situations where the Push Initiator wants to cancel a previously sent push submission. This might be done if some more up-to-date content becomes available that supersedes a previous push message before the Push Initiator has received result notification (for example, if a service has tried to notify a user of a new stock price, and then the stock price changes again before the first message was delivered). The push cancellation operation consists of a push cancellation message sent from the Push Initiator to the Push Proxy Gateway, and a push cancellation response sent back from the Push Proxy Gateway to the Push Initiator.

The status query operation has a status query message sent from the Push Initiator to the Push Proxy Gateway, and a status query response sent back to the Push Initiator from the Push Proxy Gateway. This is similar to the result-notification operation in that it is a mechanism for the Push Initiator to determine the status of a previously sent push submission; but it is different in that the Push Initiator originates it. This might be done when a Push Initiator has not yet received a result-notification message from the Push Proxy Gateway, but the Push Initiator needs to know what is happening with the push submission.

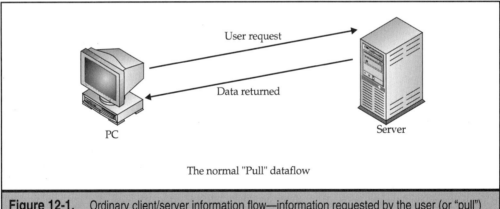

Figure 12-1. Ordinary client/server information flow—information requested by the user (or "pull")

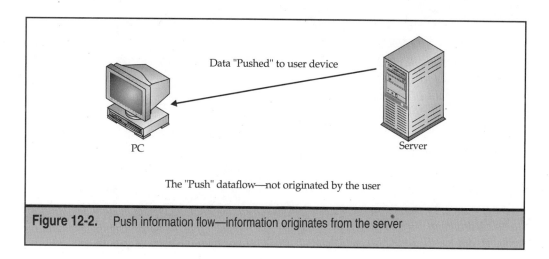

Figure 12-2. Push information flow—information originates from the server

The client-capability query operation is how a Push Initiator can find out the capabilities of a specific WAP client. A client-capability query message is sent from a Push Initiator to a Push Proxy Gateway, and the client-capability query response is sent back to the Push Initiator from the Push Proxy Gateway. The response message includes such information as the versions of WAP and WML that the device supports, the deck and message size it will accept, and the device class.

Finally, the bad-message response operation is a means for the Push Proxy Gateway to let the Push Initiator know that it could not make sense out of a previous message. If a Push Proxy Gateway is unable to determine the format of a particular message, it will send a bad-message response to the Push Initiator, which includes the content of the original bad message.

When you consider developing push-related content for mobile users, keep in mind the limitations of the typical mobile devices. Although a content provider can get information such as how large a deck a WAP device can accept (using the client-capability query operation), you need to remember that you have no idea what the user will be doing with his device at the time that content is pushed to it. It is quite possible that a user could be in the middle of some mobile application that is currently taking up most of the device's memory. If you pushed content to the device that uses the maximum deck size it can handle, then whatever the user is doing at the time of the push may get summarily dumped.

It is a good idea to push small messages that give users a chance to accept or reject the larger chunk of content that may be behind it. A message telling the user that there is a new price in a stock that she has previously asked to be tracked, and providing a link to get all the details, makes a lot more sense than just pushing all the details and possibly upsetting the user.

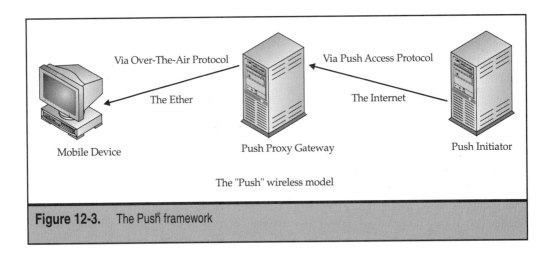

Figure 12-3. The Push framework

TELEMATICS

The passing of information from one computer to another via a telephone line or other electronic link is called *Telematics*. The basic theory is that by giving access to any form of knowledge anywhere, it will speed up the diffusion of information, save time, increase collaboration between individuals and groups, and improve the quality of decisions.

Telematics combines all the possibilities inherent in wireless voice and data communications, along with Global Positioning System (GPS) location capabilities, to deliver location-specific security, information, and productivity-enhancing services to people on the move.

Location-Sensitive Information

Telematics provides "smart" information, making information *location* sensitive and tailored to precisely where customers are and what they're doing. This can provide greatly enhanced security, navigation, and convenience to mobile consumers. Just as the Internet made information *time* sensitive in an unprecedented way by offering seamless and unlimited updatability, so Telematics makes information location sensitive as it's never been before. For the first time in history, information can be selected, modified, presented, and utilized relative to location. The impact on how we do business, on how products and services are marketed, and on how we live will be tremendous.

For example, some of the services that can be provided by Telematics are

- ▼ Emergency services location and roadside assistance. The emergency services will be able to precisely pinpoint the location of the mobile device in order to render assistance.
- ■ Vehicle tracking. Useful in instances of car or mobile device theft.

- Accident notification. When fitted to a car, the device can be used to detect air bag inflation and automatically notify the emergency services. If integrated with in-vehicle sensor systems, Telematics can send additional data to emergency services about the collision (direction of impact, speed at time of collision), thereby enhancing the specific type of rescue response.

- Travel routes. If the device always knows exactly where you are located, all you have to do is provide a destination, and the mobile device can tell you exactly how to get there by simple audio commands, or by displaying the route on the screen.

- Traffic alerts. This can either be done in conjunction with the route planning above, or separately.

- Providing the exact location of children, the elderly, infirm or handicapped persons for emergency services in order to render assistance if needed. Locking or unlocking of vehicle doors. With some kind of identification verification, the response center can control the electronic systems in the car to open the door if the owner has locked his keys in the car. This facility alone will save thousands of man-hours spent by roadside services opening cars for forgetful owners.

- Information services. This is where the marketing types can have a ball, because they can push relevant marketing or advertising data to users based on location. This is where the ability (for example) to send a user details of a special offer of a free cup of coffee if she drives in to the McDonald's restaurant just a couple of blocks ahead for a burger, becomes a major revenue-generating activity.

The actual nuts and bolts of Telematics are quite straightforward. All that is done is to integrate one of the current location-based technologies with a wireless communications system. If these technologies are then combined with a call-center response operation that can match latitude and longitude readings with various information databases (roadside assistance providers, hotels, restaurants, etc.), then you have the basic structure. The quality of the service provided is then limited only by the number and quality of additional information databases that you add to it.

Applications for Telematics

Although car manufacturers initially funded the research, the potential application of Telematics for mobile phones is virtually unlimited. With its broad functionality, it is fairly easy to predict that Telematics will appeal to a large group of consumers. Car manufacturers are already moving toward factory-installed Telematics-based emergency response systems as a standard safety feature in all their vehicles. The immediate benefits to them are that Telematics can maintain state-of-the-art product offerings, enhance customer loyalty, and enable remote diagnostics.

But that is only the beginning. It is possible to predict a future where Telematics is used in a wide variety of wireless applications, bringing personalized, web-based information to automobiles and any wireless communication. For consumers, Telematics brings added peace of mind, knowing that help can be summoned to the caller's exact location anytime it's needed, or that a loved one who is missing or overdue can be located.

For example, if a driver is involved in an emergency, he or she presses a dedicated button in the vehicle or a direct line on a handset to summon a response. The response center receives the call and the caller's location simultaneously and can immediately notify the correct emergency services. The response center can then, if necessary, help direct those services to the exact location of the car. Similarly, an injured driver can talk on the phone with a trained specialist who can assess his situation and advise him until help arrives.

The peace-of-mind factor is important to most people. For example, I am a husband and father, and I for one will certainly be providing each member of my family with such a device when it becomes commercially available, and I am sure this will be true for a large number of people.

Telematics will also bring consumers added convenience whenever they are mobile. It cuts down on time spent in a variety of vehicle-related inconveniences. Drivers can receive information about traffic conditions, the quickest route to their destination, or to a specific address or location. It can be used to find a lost or stolen car, or to check on a driver's progress if she is running late. Remote unlocking service means there's no time lost when keys are accidentally locked in the car. If a driver has car trouble, Telematics can be used to perform remote diagnostics, direct the driver to the nearest garage, or call for roadside assistance.

For wireless carriers and car manufacturers, Telematics offers an advanced way for them to better serve their customers—providing personalized service at the touch of a button. For businesses, Telematics offers the potential opportunity to communicate with mobile customers when they are in the vicinity of their business. And this is where the combined technologies of push and Telematics can make such a major difference to businesses.

PUSH AND TELEMATICS TOGETHER

Push by itself is very much a shotgun approach. It can deliver stock prices, shotgun-advertising promotions, and user-requested updates to specified information services. The user could be anywhere in a particular country. By interrogating some services, the user's location could be pinned down to within a particular radio cell. With triangulation between two or three radio masts, the user could be located to within a couple of city blocks. This is all exceedingly complex and not very accurate if someone wanted to send a marketing message to the user that is any more specific than "Eat Wheaties!"

The Telematics approach of using a Global Positioning System to locate the user can be accurate within one or two meters (three to six feet). Now businesses can *really* pinpoint their prospective customer. To return once more to the McDonald's example, you could target anyone traveling northbound on Main Street between 11:00 a.m. and 1:00 p.m. with one promotion, and have other promotions for different times of the day so you could even catch the same user coming *back* down Main Street later in the day with a different offer.

Telematics will change the way advertising and promotion are done. Changes will come slowly at first, but in a much more major way once the mobile devices are available. According to a survey done by car manufacturers that are already supplying in-car devices, 70 percent of the users say that they would like the same functionality outside of their vehicles. Currently, the GPS functionality is quite bulky, and better suited to being installed in vehicles than carried in a shirt pocket. As the sizes of the components reduce with advances in technology, it will be possible to include the same functionality in mobile devices without additional penalties.

User Privacy

One of the major problems with push, with or without Telematics, is that unregulated push content will be a worse problem than "spam" e-mail on the Internet. It is not hard to visualize a walk or drive through a downtown area becoming a nightmare of mobile devices ringing every few seconds with an offer from the store you are just passing, or the latest "not to be missed" insurance/investment/pension deal. Aside from becoming downright irritating, these messages can, more importantly, block incoming calls, which is why you are carrying the mobile device in the first place.

Mechanisms to handle this are yet to be put in place. Perhaps self-regulation by content providers will be a part of the solution. Sending an initial message asking the user to agree to receive such promotional messages, and then only sending the messages if the user has agreed, would be one method. Unfortunately, such a system is open to abuse, and we can look forward to at least a period of indiscriminate messaging until a more permanent solution is found.

SUMMARY

This chapter has covered the basics of how push technology and Telematics are supposed to work. Although all of the above is clearly specified, at the time of writing, nobody has actually done it! Crazy, huh? This will change as Telematics becomes more of a reality and therefore more attractive to corporations to invest in the software development.

CHAPTER 13

What the Future Holds

It is easy to get carried away with unrealistic predictions of technological advances. Absurd ideas are rife, and are more like science fantasy than a rational prediction—for example, "Activating his communicator by tapping his earlobe twice, the boy instructed the computer to teleport him back home." (Everybody knows that tapping an earlobe a lot would be quite painful.) However, technological advances are occurring on a daily basis, and while it is impossible to predict what new thing will suddenly be invented tomorrow, it is possible to cover the technologies that exist today and to make some calculated guesses about what can realistically be achieved.

TECHNOLOGY WITH USERS IN MIND

One thing that tends to be overlooked in all of the predictions is exactly how the individual user is likely to react to the new technology. Just as the design of a WAP application is vital to its perceived usefulness to the user, leaving the user out of the equation of technology advances can be very expensive in terms of wasted research time and dollars. The question to ask is, "Would *I* use this, if it were available now?"

The mobile phone offers a simple example. Many car accidents occur because the driver is using a mobile phone, and therefore does not have both hands properly on the steering wheel to fully control the car. This is well known, and legislation now exists in some countries to prohibit the use of a mobile phone without a hands-free kit. The trouble is that people jump into the car and either don't have a hands-free kit installed, or they forget to plug in their phones.

But with a technology called Bluetooth (which I cover in more detail in a minute), the car could be manufactured with a device preinstalled that automatically switches the mobile phone to an in-car hands-free phone, without even having to take the phone out of your pocket. The user wouldn't even have to configure the system; it would happen automatically.

This particular usage of technology makes the user's life easier and safer, and you can immediately see that this would gain acceptance without too many problems. I personally am always forgetting to plug in the phone when I get in the car, which means that as soon as the phone rings I have to find somewhere to pull over to take the call. This is not always the easiest thing to do on English country roads.

I would be willing to pay for a device that I could put in the car if it meant that I would never have to think about the problem again. What's more, as an employer I would be willing to have these devices installed in all my employees' cars so that I knew I could call them any time without worrying about their personal safety, or their ability to take the call.

It would also be nice to "talk" to a web site while I was driving so that I could get the latest news or traffic reports read to me, or send an e-mail verbally, so that I never had to take my eyes off the road. This is possible with VoiceXML, which allows me to talk to a suitably enabled web site via an ordinary phone or mobile phone, and have the data read out loud to me.

The great thing about these simple examples that I have mentioned so far is that they are not some imaginary new technology, but realistic ways of using technology that would enable me to do things that I would already like to do but am currently unable to do at this time of writing. I do a fair amount of driving, and a lot of this is "dead," or nonproductive, time. If I could have my e-mails read to me and respond to them verbally, or access other needed information while driving—yes, please!

The other point to bear in mind is that most, if not all, of this technology has been under development for some time and is just waiting for the right application to emerge so that it can join the mainstream technologies. We have had the capabilities for working from a remote location, or telecommuting, for many years now. Making it easier is nice for the people already doing it, but the increased capability still won't necessarily overcome the prejudices of many office managers and business owners who prefer to see the body in a chair in the office in front of them.

So let's take a more in-depth look at some of the technologies available right now, for developing tomorrow's applications.

BLUETOOTH — CUTTING THE CORDS

Bluetooth is basically a cable-replacement technology. Consider the current standard computer setup: you have your keyboard connected to the computer, as well as a mouse, a monitor, possibly a printer, a scanner, and so on. These are usually all connected to the PC by cables. A Bluetooth chip is designed to replace cables by taking the information normally carried by a cable and transmitting it at a special frequency to a receiver Bluetooth chip in the computer, phone, printer, or wherever.

Initially created by Ericsson, Bluetooth is a standard for a small, cheap radio chip to be plugged into computers, printers, mobile phones, and so on. That was the original idea, but it soon became clear that a lot more was possible. You can transmit information between *any* two devices: computer and printer, keyboard and mobile phone, and so on. The projected low cost of a Bluetooth chip (around $5) and its low power consumption, means you can literally place one anywhere.

You could have Bluetooth chips in freight containers to identify cargo when a truck passes through customs or drives into a storage depot; or a headset that communicates with a mobile phone in your pocket or in another room; or an e-mail that has been sent to your mobile device being printed out as soon as you get within range of the office computer.

I currently use a wireless keyboard and mouse, and I notice many conveniences. I can sit back in my chair with my keyboard on my lap and continue to type without having the keyboard cable drag half the papers on my desk onto the floor. I can use the mouse without that really annoying "drag" of the mouse cable when it gets to its limit, or gets snagged on the edge of the desk or something. I can, and do, take the keyboard across to a desk on the other side of the room to look up a reference, and just type it from there and then come back to my own desk with my keyboard. What I have typed (or mistyped!) is there on the screen waiting for me to continue.

Cables have become the bane of many offices and home offices. Most of us have experienced the pleasures of trying to figure out what cable goes where, and getting tangled up in the snarls behind the desk. With two sons and my wife running her own business from home, we have managed to "grow" our own five-PC network, with cables running all over the house. With Bluetooth, all this would change.

Bluetooth also provides a global standard for wireless connectivity, and I have been told that a Bluetooth microchip will be built into virtually all digital devices manufactured from the year 2002 on. The way it works is that when two Bluetooth-equipped devices come within about 10 meters (30 feet) of each other, they can establish a connection. And because Bluetooth uses a radio-based link, the chips don't require a line-of-sight connection in order to communicate. Your laptop could send information to a printer in the next room, or your microwave oven could send a message to your mobile phone or PC telling you that your meal is ready.

The radio chip operates in a globally available frequency band of 2.4GHz, ensuring compatibility worldwide. The Bluetooth technologies make all of the connections instantly and provide fast and secure transmission of both voice and data.

Here are some of the simpler things that could be brought about as a result of using Bluetooth:

- A "personal area network," allowing seamless integration of computing or mobile resources in your car with other communication and computing platforms at home and at work
- Automatic switchover between a hands-free car phone and a mobile device
- Wireless updates of all to-do lists, contact lists, and so on, as soon as you get within range of your computer
- Synchronization with local companies and service providers for information "push" and e-commerce
- Secure access to private data networks, including your office e-mail systems
- "Over-the-air" diagnostics in your car of the engine management system and reprogramming if necessary

In the future, Bluetooth is likely to be standard in tens of millions of mobile phones, PCs, laptops, and a whole range of other electronic devices. As a result, there is going to be a lot of demand for new innovative applications, value-added services, end-to-end solutions, and much more. The possibilities are limitless.

At first, Bluetooth will simply begin to replace the cables that connect various digital devices, and as the number of connections increases, so will the potential. As more and more manufacturers adopt Bluetooth and create devices that support it, developers will find new, previously unimagined ways of applying its power.

Bluetooth is one of the key technologies that can make the mobile information society possible, blurring the boundaries between home, the office, and the outside world. And in case you were wondering why it is called Bluetooth, I can tell you that it is named after a Danish Viking and king, Harald Blåtand (or Bluetooth in English), who lived in the late tenth century. Harald united and controlled Denmark and Norway, hence the selection of the name—uniting devices through Bluetooth. Apparently, Harald had a great liking for blueberries—so much so that his teeth were stained blue, leaving him with a rather distinctive set of choppers.

VOICEXML—A NEW SLANT ON "WALKIE/TALKIE"

The most natural way for people to communicate is by the spoken word. Because something like one in five people in the Western world are functionally illiterate (a horrendous figure, but true), many people will never in their day-to-day lives be able to physically interact with a PC. But they can talk and listen.

So, what would happen if you could use an ordinary mobile phone to access the Internet, talk to it like you would talk to a person, and have it talk back to you with the information that you asked for? This is the premise behind VoiceXML, which is a set of standards, publicly released in March 2001, for the acceptance of voice commands and the details of what to do in response to those commands. It is constructed as a set of tags in XML fashion, with which you can prompt the user for information and accept spoken commands from the user.

Probably the easiest way to explain VoiceXML is to give a short example of the code:

```
<?xml version="1.0"?>
<vxml version="1.0">
<form id="ChooseProductType">
 <field name="document.generic.ProductGroup">
  <grammar>
   <![CDATA[
          [
          [ (dtmf-1) book books] {<option "BOOKS">}
          [ (dtmf-2) music cd cds] {<option "MUSIC">}
          [ (dtmf-3) video videos movie dvd] {<option "VIDEO">}
          ]
   ]]>
  </grammar>
  <prompt>
   <audio>Please select a product type. Choose books, music or video now.</audio>
   <pause>2000</pause>
  </prompt>
  <filled>
   <result name="BOOKS">
     <audio>You selected books.</audio>
     <goto next="#ChooseBooksType"/>
   </result>
   <result name="MUSIC">
     <audio>You selected music.</audio>
     <goto next="#ChooseMusicType"/>
```

```
      </result>
      <result name="VIDEO">
        <audio>You selected video.</audio>
        <goto next="#ChooseVideoType"/>
      </result>
</form>
</vxml>
```

In this code, the application speaks to the user, "Please select a product type. Choose books, music or video now." If the user says, "Video," or "videos," or "movie," or "DVD," she is taken to the result, named "VIDEO." The user is then told, "You selected video," and the application takes her on to a form named "ChooseVideoType," where the choice can be narrowed some more.

With what you have already learned from this book, you can see that you should be able to pick up the VoiceXML syntax quite easily and therefore expand your own ability to create more versatile applications.

Because we are talking here about accessing the Internet from an ordinary phone as well as a mobile device, there are no additional costs that need to be incurred for users to access Internet services via voice portals. Do you get what I'm saying here? VoiceXML allows people with no computer, no PDA, not even a mobile phone, to access the Internet using the most natural medium of all—their voice. This opens up the world (via VoiceXML applications) to anyone with a phone. This is exciting stuff.

A phone also allows eyes-free and hands-free operation, enabling the use of the Internet in situations where using a device with a screen is not optimum, like the car, for example.

Since VoiceXML avoids the costs associated with buying and activating new wireless devices and services by using an existing infrastructure and services, there is plenty of room for the rapid growth of "voice portals," which is the name given to this type of service.

TELEMATICS—WE KNOW WHERE YOU ARE

The topic of Telematics is covered in the previous chapter, but it is worth mentioning here in order to see just how it might fit into the scheme of things when it is combined with the other technologies of VoiceXML and Bluetooth. With location based services you have the ability to talk directly to your customer or user in a way that was previously impossible. You know that the user is walking down Main Street, or stationary at the corner of Main Street and High Street. You know that the user is heading towards a bookstore, and that this particular user has also expressed a desire to be notified of any special deals on bestsellers that might be occurring from that chain of stores.

If all this data is put together correctly, you could well find that the bookstore has just made another sale as a result of your application.

In a more practical example, you or one of your family get lost walking in the woods, or run in to difficulties while sailing just offshore, or have a car breakdown in a dangerous neighbourhood. The emergency services can pinpoint you and get you the assistance you need with minimum delay.

For the reasons given in the last example alone, mobile operators in the United States will be legally obliged (by 2002) to provide the location, within a few meters, of any mobile device placing a call. While this is ostensibly to assist emergency services, it does raise serious issues of privacy and the "right to know" for more general calls.

BRINGING IT ALL TOGETHER

Bluetooth, VoiceXML, and Telematics, along with WAP, are all separate technologies, with different protocols, standards, specifications, and guiding bodies. Any one of these technologies is capable of achieving wonderful things by itself, and each one breaks new ground in terms of capabilities. But where the future gets really exciting is when we start to look at how these technologies can be used together. Let's face it, VoiceXML might be really exciting because it opens up the availability of the Internet to anyone with just a phone. But it is more likely that VoiceXML will be used when hands-free operation is needed, and other more visual methods used when possible.

Imagine some of these everyday examples: you board a bus or an airplane and your fare is automatically paid by your mobile phone and added to your phone bill. Or you get an automatic text message letting you know that your children have arrived home safely from school. Or you write e-mails on your laptop or PDA on the airplane, and then when you land and switch on your mobile device, your phone can automatically send the messages via the Internet.

Another use that springs to mind is in the field of robotics. The combination of VoiceXML and Bluetooth means that all of the voice recognition and logic systems, which are processor intensive and memory hungry, can be handled at the server, and only the instructions need to be sent to the robot itself. This means that the robot can be a much simpler mechanism, and therefore much cheaper to manufacture. It doesn't have to be even vaguely sophisticated, as all of the intelligence can be off-loaded to the server. This means that the robot can be nothing but a set of motors with a controller that is linked via Bluetooth to the server—exactly right for those house-cleaning robots that my wife has always wanted.

Other applications may sound familiar, but the way they are presented with the above technologies is what is going to make the next "killer app" that takes the world by storm. The applications are

- Customized information and entertainment on demand
- News, financials, weather, and sports
- Audio books and e-books
- Music
- Online games
- Voice-activated Web searches and information retrieval
- E-commerce

When applications get a little more sophisticated, we will be able to book a flight, or a theater ticket, or a rental car simply by verbally telling the phone or mobile device to book it for us. Communication with the office or our customers will be an integral part of our lives, with telecommuting being a reality from anywhere. We will be able to make decisions based on current and up-to-date data, on any topic and at any time, and from any location.

Entertainment, while at home, while traveling, or while at work, will be as natural as breathing, with the ability to select the exact music we want.

With Bluetooth, control systems for the house and office can and will be routine. The ability to control functions from our mobile devices will be second nature.

And how about having total connectivity on the road? We may have the ability to create e-mails (or voice mails) from the car, or to compose letters or build spreadsheets, all while driving to appointments that we are being guided to via the Telematics system. And then we will have these same documents uploaded to our PCs when we pull into the parking lot or driveway so that we can print them with a simple verbal command. These are the things that will make our lives that much more productive.

I usually get my best ideas when I am in the shower in the morning. The only trouble is, I don't have a pen and paper in the shower with me. Wouldn't it be great to send myself a memo from the shower, or to dictate pseudocode while in the car for the next great application, or even write a book while stuck in rush hour traffic?

The next generation of data transmission will be faster, and then faster still. The underlying protocols, however, will be the same. How we, as developers, can use these to create better solutions for ordinary users will come down to our ability to spot the problems that they may be having and to our creativity in building workable applications that truly solve those problems.

Technology will always be improving. As I write this, I have just been reading an article from IBM that states that they have found a way to build transistors one hundredth of the size of those currently being laid down in silicon chips. This means either one of two things: the circuits get smaller and lighter, or alternatively, you can cram in 100 times the capacity in a device the same size.

This is today, and this is real, not fantasy.

What does tomorrow hold?

Watch this space.

CHAPTER 14

WMLScript Reference

Although I have already covered WMLScript when it was first introduced in the book, there will be times when you will only be interested in looking up a specific piece of syntax or command. This chapter is included for you to refer to when those situations arise.

CASE SENSITIVITY

WMLScript is a case-sensitive language. All keywords, variables, and function names must be written consistently in terms of case in order to be recognized; otherwise you will introduce all sorts of strange bugs into your code. For example, "dale", "Dale", and "DALE" are all considered to be different entities by WMLScript.

There are several accepted conventions specifying when and how to capitalize your words. Some people like to write in all lowercase letters, regardless. Others insist that the first letter of each whole word should be capitalized, as in "GetLetter" or "InsertSpace". In Java, the convention is to have the first word in lowercase and the second word capitalized, as in "getLetter" and "insertSpace".

The bottom line is that it really doesn't matter which convention you choose to follow, as long as you are consistent with the one you pick.

WHITESPACE AND LINE BREAKS

WMLScript ignores spaces, tabs, and newlines that appear in your script, except those that are included in string literals. The WMLScript compiler recognizes all of the following strings as being different:

```
"The cat sat on the mat."
"Thecatsatonthemat"
"Thecatsat    onthemat"
```

The WMLScript compiler recognizes the following three commands as being the same, as the whitespace is ignored in commands:

```
function helloworld()

function           helloworld()

function
      helloworld()
```

COMMENTS

There are two different types of comments in WMLScript, as in several other programming languages.

Line comments are single-line comments that start with two slashes (//) and end at the end of the line. The slashes can also be added to the end of a working line of code to clarify a point:

```
X = "451"      // the temperature in Fahrenheit at which paper ignites
```

Block comments are multiple lines that start with a slash and an asterisk (/*) and end with the reverse (*/). These are useful if you have lengthy comments that run on for several lines. As the general principle is that more comments are better than fewer, you will probably be making extensive use of this type of comment. Here is an example:

```
/*
Function InsertSpace(string, x)
 This function inserts a space into a string at a specified position.
 Syntax: InsertSpace(string, x);
 Parameters:
 string = the text string
      x = character position (zero based) where the space is inserted
 Example: InsertSpace("thecat",3);
 Result: "the cat".
*/
```

Block comments cannot be nested, as the first instance of "*/" will terminate the comment, and whatever is after that will be looked at as an attempt at a command. If what follows the first "*/" is supposed to be more comments, then you will get either errors or spurious commands being acted on. To illustrate:

```
/* This is the first block comment
This function tells us the time in London regardless of time zone

/* The Syntax is:         // Nested block comment...
Function LondonTime()
*/    // This line will end all block comments

It is a good function to use    // This line will generate an error
*/    // End of first block comment
```

CONSTANTS

Constants are also known as literals—these are values that are hard-coded inside your script. There are several different types of constants: integer, floating-point, string Boolean, and invalid constants.

Integer Constants

Integer constants are whole numbers that are expressed as ordinary decimal (base 10), hexadecimal (base 16), or octal (base 8) numbers.

- **Decimal**   A decimal integer literal is a string of digits that does not begin with zero.
- **Hexadecimal**   A hexadecimal integer literal is a string of digits that begins with 0X or 0x. Hexadecimal integer literals can contain the numbers 0–9 and the letters a–f or A–F.
- **Octal**   An octal integer literal begins with zero. Octal integer literals contain only the numbers 0–7.

Here are a few examples:

```
45         // Decimal
0x2D       // Hexadecimal
055        // Octal
```

Take care when you type a leading zero, as any numbers following it will be interpreted as octal. Also note that the written numbers 7, 0x7, and 07 are identical, and are interpreted as 7, no matter which convention you use. This can cause unexpected results if you are not careful.

Floating-point Constants

Floating-point constants are numbers that contain decimal places. A floating-point constant can contain both decimals and exponents. For a constant to be considered floating-point, it must have at least one digit and either a decimal point or an exponent.

A floating-point constant can contain any of the following:

- A decimal integer (for example, 12)
- A decimal point (for example, .14142)
- A fraction (for example, 1/2)
- An exponent (for example, e2); an exponent is indicated by an e or E followed with an integer

As a result, the value 3.14 can be represented in any of the following ways:

- 3.14
- 3.14e0
- 3.14E0
- .314E1
- 314e-2

String Constants

A string constant is any sequence of zero or more characters enclosed within double quotes ("") or single quotes (''). The beginning and ending quote characters must be the same. You cannot start a string with a single quote and end it with a double quote, or vice versa.

Here are some example string constants:

```
'Click Here to Continue'
"Welcome to Shoppers Hell!"
"Created on 7th June, 2001"
```

Some characters, such as a form feed, cannot be represented within string constants, but WMLScript supports special escape sequences by which these characters can be represented. Table 14-1 shows the symbol sequence to use for each special character.

Character	Symbol Sequence
Apostrophe or single quote (')	\'
Double quote (")	\"
Backslash (\)	\\
Forward slash (/)	\/
Backspace	\b
Form feed	\f
Newline	\n
Carriage return	\r
Horizontal tab	\t

Table 14-1. Escape Sequences for Special Characters

Character	Symbol Sequence
The character with the encoding specified by two hexadecimal digits *hh* (Latin-1 ISO8859-1)	\x*hh*
The character with the encoding specified by the three octal digits *ooo* (Latin-1 ISO8859-1)	*ooo*
The Unicode character with the encoding specified by the four hexadecimal digits *hhhh*.	\u*hhhh*

Table 14-1. Escape Sequences for Special Characters *(continued)*

Boolean Variables

Boolean variables are variables that store values of *true* or *false*.

Invalid Variables

The Invalid variable is a variable that has a type of "Invalid". They are not so much variables as error-value comparators. This is useful for trapping errors that you might otherwise overlook.

Return Value

Invalid

Example

As the following will give us a "divide by zero", an error is generated and the variable is converted to the type **invalid** which can then be tested for.

```
var x = 8;

var y = 0;

if ((x/y) == invalid) {

display error message

};
```

RESERVED WORDS

Reserved words have a special meaning in WMLScript programs and therefore can't be used as identifiers. The reserved words in WMLScript are listed in Table 14-2.

access	agent
break	case
catch	class
const	continue
debugger	default
delete	div
do	domain
else	enum
equiv	export
extends	extern
finally	for
function	header
http	if
import	in
isvalid	lib
meta	name
new	null
path	private
public	return
sizeof	struct
super	switch
this	throw
try	typeof
url	use
user	var
void	while
with	

Table 14-2. WMLScript Reserved Words

VARIABLES

Variables are symbolic names for values coded in your script. You use variables to store and manipulate temporary program data.

WMLScript requires you to declare variables before using them for the first time. You can initialize variables with their starting values when they are declared. If no initial values are specified, then the variables are (theoretically) automatically initialized to contain an empty string value ("").

I say "theoretically," as it is never good practice to leave such things to the compiler. If you get into the habit of assuming the compiler will do certain things, you may well find yourself handling some strange bugs when you start coding in other languages.

Variable Declaration

The `var` keyword is used to declare a variable. More than one variable may be declared in the same var statement, but they must be separated with commas. A variable declaration is terminated with a semi-colon (;). Variable names within a function must be unique.

Here are a few examples:

```
var a;
var x, y, z;
var size = 3;
```

Variable Scope and Lifetime

A variable's *scope* is the part of the program in which the variable is "visible" to the rest of the code. A variable's *lifetime* is the time between the variable's declaration and the end of the function in which it was declared.

In the following example, the variable `newPrice` is declared within the function `priceCheck()`. After the function has finished, the variable `newPrice` will not exist any more.

```
function priceCheck(givenPrice) {    var newPrice = givenPrice;
if (newPrice > 100) {

}
else {
    newPrice = 100;
};
return newPrice;
};
```

A variable is accessible only within the function in which it has been declared.

DATA TYPES

WMLScript supports different data types internally. You do not have to specify the variable type, and any variable can contain any type of data at any given time. This is like the early versions of Basic, and quite unlike languages like Visual C++, where the type of every variable has to be declared when the variable itself is declared.

What this means is that in WMLScript you can declare variables as easily as this:

```
var flag = true
var wholenumber = 49
var roomtemp = 21.1
var myname = "Dale Bulbrook"
var exception = invalid
```

This automatic variable data typing actually makes the programmer's job easier in many ways, but it can create a problem if the conversion from one type to another doesn't work the way you expect. Such problems are very rare.

The only things that you really need to know about the variable types are the limits that you have to work with for each type. These limits are outlined in Table 14-3.

One disadvantage of not being able to declare the variable type is that the system will never second-guess your calculations and will assume everything is working exactly the way you intended. With variable typing, the compiler can spot and mark as an error any attempt to assign a value to a "wrong" or different type.

In C++, for example, if you say this,

```
Var x as integer;
Var y as string;
Var z as string;
z = x + y;
```

you will get an error message at compile time telling you that you are trying to perform arithmetic with a string data type, and you can see exactly what you have done incorrectly.

Type	Size
Integer	−2147483648 to 2147483647
Floating-point	1.17549435E−38 to 3.40282347E+38.
String	N/A
Boolean	N/A
Invalid	N/A

Table 14-3. Size Limits for WMLScript Data Types

In WMLScript however, you can say this,

```
Var x = 1;
Var y = "one";
Var z = x + y;
```

and you will get no error at all. In fact, the variable z will contain "1one". The conversion from 1 as a number to "1" as a string is done automatically when the addition is performed. Why is the number 1 converted to a string? Because it is a more logical assumption for the system to make than converting the string "one" to a number. Neat though this may be, you may well find that "1one" is not always the result that you wanted.

There are a whole set of rules that exist to determine how the conversions are done, in what sequence, and so on. You can find the full rules at www.wapforum.org if you need them.

PRAGMAS

A *pragma* is a *directive* that generates special behavior from the compiler. You should specify any pragmas at the beginning of the file, before declaring any functions.

All pragma directives start with the keyword use and are followed by specific attributes. The following pragma directives are permitted, and will be looked at individually:

- use url
- use access
- use meta

External Files

You can use a URL to access a WMLScript file, and you can use a use url pragma to bind the external file to the WML file, as shown here:

```
use url UtilityScript "http://www.host.com/app/script";
```

To call an external file, you must specify the following:

- The URL of the WMLScript resource (in this example http://www.host.com/app/script)
- The resource's name (in this example UtilityScript)

Then, inside the function declarations, you can use just the resource name that you declared in the use url pragma. To use the same example:

```
function test (par1, par2) {
return UtilityScript #check(par1-par2);
};
```

The following occurs in the example function call above:

- ▼ The pragma specifies the URL to the WMLScript file.
- ■ The function call loads the file by using the given URL.
- ▲ The file's content is verified and the specified function (in this case the function `check()`) is executed.

The `use url` pragma has its own namespace for local names. However, the local names must be unique within any given file. WMLScript 1.1 supports URLs and relative URLs without a hash mark (#) or a fragment identifier. For relative URLs, the base URL is the URL that identifies the current file.

The specified URL must be escaped according to the URL-escaping rules. No compile-time automatic escaping, URL syntax, or URL validity checking is performed.

Access Control

You can use an access-control pragma to protect a file's content from unauthorized access. You must call the access-control pragma before calling external functions, and having more than one access-control pragma in a file will generate a compiler error.

Every time an external function is invoked, the compiler performs an access-control check to determine whether the destination file allows access from the caller. The access-control pragma specifies domain and path attributes against which the access-control checks are performed. If a file has a domain or path attribute, the referring file's URL must match the values of the attributes. Matching is done as follows:

- ▼ The access domain is suffix-matched against the domain-name portion of the referring URL.
- ▲ The access path is prefix-matched against the pathname portion of the referring URL.

Domain and path attributes follow the URL capitalization rules.

Domain suffix matching is done using the entire element of each subdomain and must match each element exactly. For example, www.wapforum.org matches wapforum.org but not forum.org.

Path prefix matching is done using entire path elements and must match each element exactly. For example, /X/Y matches /X, but not /XZ.

The domain attribute defaults to the current file's domain. The path attribute defaults to the value " / ".

To simplify the development of applications that may not know the absolute path to the current file, the path attribute accepts relative URLs. The user agent converts the relative path to an absolute path and then performs prefix matching against the path attribute. For example, if the access-control attributes for a file are these,

```
use access domain "wapforum.org" path "/finance";
```

the following referring URLs would be allowed to call the external functions specified in this file.

- http://wapforum.org/finance/money.cgi
- https://www.wapforum.org/finance/markets.cgi
- http://www.wapforum.org/finance/demos/packages.cgi?x+123&y+456

The following URLs would *not* be allowed to call the external function.

- http://www.test.net/finance
- http://www.wapforum.org/internal/foo.wml

By default, access control is disabled.

Metadata

Metapragmas specify a file's *property name* and *content*; metapragmas can also specify a file's *scheme*, which specifies a form or structure that may be used to interpret the *property name* value—the values vary depending on the type of metadata. The attribute values are string literals.

Metapragmas do not define any properties, nor do they define how user agents must interpret metadata. User agents are not required to act on the metadata.

Metapragmas can have the following attributes:

- **Name**   Name specifies metadata used by the origin servers. The user agent should ignore any metadata named with this attribute. Network servers should not emit WMLScript 1.1 content containing name metapragmas.

  ```
  use meta name "Created" "18-June-1999";
  ```

- **HTTP equiv**   HTTP equiv specifies metadata that indicates that the property should be interpreted as an HTTP header. Metadata named with this attribute should be converted to a wireless session protocol (WSP) or HTTP response header if the file is compiled before it arrives at the user agent.

  ```
  use meta http equiv "Keywords" "Script, Language";
  ```

- **User agent**   User agent specifies metadata that is intended to be used by the user agents. This metadata must be delivered to the user agent and must not be removed by any network intermediary.

  ```
  use meta user agent "Type" "Test";
  ```

OPERATORS

An operator is a symbol or word, such as **+** and **Or**, that indicates an operation to be performed on one or more elements. There are several classes of operators, including assignment, arithmetic, logical, string, and comparison operators.

Assignment Operators

An assignment operator is a character that assigns a value to a variable. The simplest form of assigning an operator is the regular assignment (=), used here:

```
var a = "abc";
var b = a;
b = "def";
```

Assignments can be made with any of the operators listed in Table 14-4.

Operator	Description
=	Assign
+=	Add (numbers) or concatenate (strings) and assign
-=	Subtract and assign
*=	Multiply and assign
/=	Divide and assign
div=	Divide (integer division) and assign
%=	Remainder (the sign of the result equals the sign of the dividend) and assign
<<=	Bitwise left shift and assign
>>=	Bitwise right shift with sign and assign
>>>=	Bitwise right shift, zero fill, and assign
&=	Bitwise AND and assign
^=	Bitwise XOR and assign
|=	Bitwise OR and assign

Table 14-4. Assignment Operators

All of the operators in Table 14-4 are shortcuts to the "standard" notation, and can all be written out the "long" way. For example:

```
a += b can be written as a = a + b   a *= b can be written as a = a * b
```

If you are ever in doubt about the visual clarity of your code, write it the long way. It won't make any difference to the speed at which your code runs, and you may well thank yourself in six months' time when you have to maintain the code.

Arithmetic Operators

Arithmetic operators are the basic binary characters that perform arithmetic operations. The following are some examples of arithmetic operators:

```
var y = 1/3;          //  (y = 0.333)
var x = (y * 3) + (y - b);
```

WMLScript supports all of the basic binary arithmetic operations listed in Table 14-5:

Operator	Operation
+	Add (numbers) or concatenate (strings)
-	Subtract
*	Multiply
/	Divide
div	Integer division
%	Remainder, the sign of the result equals the sign of the dividend
<<	Bitwise left shift
>>	Bitwise right shift and sign
>>>	Bitwise right shift with zero fill
&	Bitwise AND
|	Bitwise OR
^	Bitwise XOR
+	Plus
-	Minus

Table 14-5. WMLScript Arithmetic Operators

Operator	Operation
--	Pre or post decrement
++	Pre or post increment
~	Bitwise NOT

Table 14-5. WMLScript Arithmetic Operators *(continued)*

Logical Operators

Logical operators combine or negate relational expressions. The basic logical operations are AND, OR, and NOT. The logical operators are listed in Table 14-6.

The logical AND operator evaluates the first operand and tests the result. If the result is false or invalid, the result of the operation is false or invalid, and the second operand is not evaluated. If the first operand is true, the result of the operation is the result of the second operand.

Similarly, the logical OR evaluates the first operand and tests the result. If the result is true or invalid, the result of the operation is true or invalid, and the second operand is not evaluated. If the first operand is false, the result of the operation is the result of the second operand.

```
weAgree = (iAmRight && youAreRight) || (!iAmRight && !youAreRight);
```

In this example, if `iAmRight` AND `youAreRight` are both `True`, then `weAgree` is set to `True` and the next line of code is executed. However, if either `iAmRight` or `youAreRight` are `False`, then the second operand is evaluated to see if both variables are `False`. If they are both `False`, then `weAgree` is set to `True`. And if both operands are false, then `weAgree` is set to `False`.

Operator	Description
&&	AND
||	OR
!	NOT (unary)

Table 14-6. WMLScript Logical Operators

WMLScript 1.1 requires a Boolean value for logical operations. Automatic conversions from other types to Boolean types are supported.

Note that if the value of the first operand for logical AND or OR is invalid, the second operand is not evaluated, and the result of the operand is invalid. For example:

```
var a = (1/0) || foo();    // result: invalid, no call to foo()
var b = true || (1/0);     // true
var c = false || (1/0);    // invalid
```

String Operators

String operators are characters that concatenate string variables. These are some examples of string operators:

```
var str = "Beginning" + "End";
var chr = String.charAt(str,10) // chr = "E"
```

The + and += operators concatenate the strings. Other string operations are supported by the standard String library (see the "String Library" section later in this chapter).

Comparison Operators

Comparison operators are characters that compare values within a variable. They are also known as relational operators. Here are some examples of comparison operators:

```
var res = (myAmount > yourAmount);
var val = (( 1/0) == invalid); // val = invalid
```

WMLScript 1.1 supports the comparison operations listed in Table 14-7.

Operator	Description
<	Less than
<=	Less than or equal
==	Equal
>=	Greater than or equal
>	Greater than
!=	Inequality

Table 14-7. WMLScript Comparison Operators

Comparison operators use the following rules:

- ▼ **Boolean**   True is larger than false.
- **Integer**   Comparison is based on the given integer values.
- **Floating-point**   Comparison is based on the given floating-point values.
- **String**   Comparison is based on the order of character codes of the given string values. Character codes are defined by the character set supported by the WMLScript 1.1 interpreter.
- ▲ **Invalid**   If at least one of the operands is invalid, the result of the comparison is invalid.

Comma Operator

Comma operators combine multiple evaluations into one expression. Commas used in a function call to separate parameters and to separate multiple variable declarations are not comma operators. The following is an example of a comma used in the `for` loop statement to separate multiple declarations:

```
for (a=1, b=100; a < 10; a++,b++) {
// do something
};
```

This initializes variable a as 1 and variable b as 100. The second parameter sets the limit to be checked on each loop iteration, in this case if a is less than 10 then it is ok to continue the iterations, and the third parameter increments varables a and b by 1 for each iteration of the loop.

Here are some examples of using the comma to separate variable declarations:

```
var a = 2;
var b = 3, c = 3;
```

Comma operators are quite different. The result of the comma operator is always the value of the last operand in the given expression.

For example:

```
var x;
var y=0;
var z;

x = (1+1), (2+2);
//x will hold the result of "4"

z = y++, x*x;
// y will be incremented by 1, but z will now hold the result of 16
```

Conditional Operator

Conditional operators assign a value to an expression based on the Boolean result of an initial statement. The following is an example of a conditional operator:

```
myResult = flag ? "Off" : "On (value=" + level + ")";
```

Conditional operators (?:) are essentially if-then statements. They take three operands, which are arranged as follows:

```
d = operand1 ? operand2 : operand3
```

In this statement, operand1 is the condition being evaluated. If the condition is true, expression d results in the value or result of operand2. If the condition is false or invalid, expression d results in the value or result of operand3.

typeof Operator

The typeof operator returns an integer value that describes the type of the given expression. The following is an example of the use of the typeof operator:

```
var str = "123";
var mytype = typeof str; // mytype = 2
```

The typeof operator does not convert the result from one type to another. It returns a return code indicating the type of data in the variable. The return codes are listed in Table 14-8.

Data Type	Return Code
Integer	0
Floating-point	1
String	2
Boolean	3
Invalid	4

Table 14-8. typeof Operator Return Codes

isvalid Operator

The `isvalid` operator checks the validity of the given expression. The following are examples using the `isvalid` operator:

```
var str = "123";
var ok = isvalid str; // true
var tst = isvalid (1/0); // false
```

The `isvalid` operator returns a Boolean value of false if the type of the expression is invalid, otherwise true is returned. `isvalid` does not convert the result from one type to another.

EXPRESSIONS

An expression is any combination of operators, constants, literal values, functions, or variable names. Expressions can be the single value of a constant or a variable, such as these:

- 567
- 66.77
- "This is too simple"
- 'This works too'
- true
- myAccount

Expressions that are more complex can be defined by using simple expressions, operators, and function calls.

- myAccount + 3
- (a + b) / 3
- initialValue + nextValue(myValues);

WMLScript 1.1 supports most of the expressions common to other programming languages.

FUNCTIONS

A *function* is a part of a program that performs a specific task, and that can be executed from more than one place within the program, and as many times as necessary or wanted. A function can have input parameters and output parameters. The input parameters are those data necessary for the function to be able to perform its assigned task, and the output parameter is the result obtained by the executed function.

Function Declarations

In WMLScript 1.1, function declarations declare a function name with optional parameters and a block statement that is executed when the function is called. All functions have the following characteristics:

- ▼ Function declarations cannot be nested.
- ■ Function names must be unique within one file.
- ■ All parameters to functions are passed by value.
- ■ Function calls must pass exactly the same number of arguments to the called function as specified in the function declaration.
- ■ Function parameters behave like local variables that have been initialized before the function body (block of statements) is executed.
- ▲ A function always returns a value. By default, it is an empty string (""). A `return` statement can be used to specify other return values.

You can use the `extern` keyword to make a function available to an outside file, as follows:

```
extern function testIt() {
var USD = 10;
    var FIM = currencyConverter (USD, 5.3);
};
function currencyConverter(currency, exchangeRate) {
    return currency*exchangeRate;
};
```

The second function in this example, `currencyConverter()`, cannot be called externally, but can only be called by functions within the same file, or in this case, the function `testIt()`.

Function Calls

Function calls return a value, such as the result of a calculation or comparison between two operands. The way a function is called depends on where the function you want to call is declared. The following sections describe three function calls supported by WMLScript 1.1:

- Local script functions
- External functions
- Library functions

Local Script Function Calls

A local script function is one that is declared and called in the same file. The following is an example of a local script function call:

```
function test2 (param) {
return test1 (param+1);
};
function test1 (val) {
   return val*val;
};
```

Local script functions can be called simply by providing the function name and a comma-separated list of arguments. The number of arguments must match the number of parameters accepted by the function.

A local script function can be called before it has been declared.

External Function Calls

An external function is one that is declared in an external file. The following is an example of an external function call:

```
// This is the first WMLScript file

use url UtilScript "http://www.this.com/utilscript.wmls";
function test3(param) {
    return UtilScript#test2(param+1);
};
// This is the file script.WMLScript

extern function test2(param) {
   // Test Number passed
   if (param > 100) {
      return true;
   }
return false;
}
```

The external function call must be prefixed with the name of the external file, and you must use the use url pragma to specify the external file. This pragma maps the external unit to a name that can be used within the function declaration. This name and the hash symbol (#) are used to prefix the standard function call syntax.

Library Function Calls

A library function call is one that calls a WMLScript 1.1 standard library function. The following is an example of a library function call:

```
function test4(param) {
return Float.sqrt(Lang.abs(param)+1);
};
```

You can call a library function by prefixing the function name with the name of the library and the dot symbol (.). In the preceding example, the function abs is prefixed by library Lang.

STATEMENTS

A statement is a syntactically complete unit that expresses one specific kind of operation, declaration, or definition. Virtually every line in a program is a statement of one kind or another.

Empty Statements

Empty statements are used when a statement is needed but no operation is required. The following is an example of an empty statement:

```
while (!poll(device));    // Wait until poll() is true
```

Expression Statements

Expression statements assign values to variables, calculate mathematical expressions, make function calls, and so on. The following are examples of expression statements:

```
str = "Hey" + yourName;
val3 = prevVal + 4;
counter++;
myValue1 = counter, myValue2 = val3;
alert("Watch out!");
retVal = 16*Lang.max(val3, counter);
```

Block Statements

Block statements (also known as *compound* statements) are a collection of statements enclosed in curly brackets that are treated as a single statement. The following is an example of a block statement:

```
{
var i = 0;
```

```
        var x = Lang.abs(b);
        popUp("Remember!");
}
```

A block statement can be used anywhere a single statement is used.

Variable Statements

Variable statements are used to initialize variables. The following are examples of variable statements:

```
function count(str) {
var result = 0; // Initialized once
    while(str != "") {
        var ind = 0; // Initialized every time
        // modify string
    };
    return result
};

function example(param) {
    var a = 0;
    if (param > a) {
        var b = a+1; // Variables a and b can be used
    } else {
        var c = a+2; // Variables a, b, and c can be used
    };
    return a: // Variables a, b, and c are accessible
};
```

Variable names, which can be any legal identifier, must be unique within a single function. The scope of the declared variable is the rest of the current function; the expression (in the example above, "=0", "=a+1", "=a+2") is evaluated every time the variable statement is executed. Variables can be initialized either to a specified value or, by default, to an empty string ("").

If Statements

An `if` statement presents a condition and one or two statements that are executed depending on the Boolean value of the condition. The following is an example of an `if` statement:

```
if (sunShines) {
myDay = "Good"
    goodDays++;
```

```
} else
    myDay = "Oh well...";
}
```

An `if` statement consists of a condition and one or two statements (or block statements). The first statement is executed if the specified condition is true; if the condition is false or invalid, the second (optional) statement is executed. The statements can be any WMLScript 1.1 statements, including other nested `if` statements.

While Statements

A `while` statement creates a loop that evaluates an expression and, if it is true, executes one or more expression statements. The following is an example of a `while` statement:

```
var counter = 0;
var total = 0;
while (counter < 3) {
    counter++;
    total += c;
};
```

The loop repeats as long as the specified condition is true.

For Statements

A `for` statement creates a loop that executes as long as a stated condition is true. The following is an example of a `for` statement:

```
for (var index = 0; index < 100; index++) {
count += index;
    myFunc(count);
};
```

The `for` statement consists of three optional expressions enclosed in parentheses and separated by semicolons, followed by a statement-executed loop.

Typically, the first expression (`var index = 0`, in the preceding example) is used to initialize a counter variable. This expression can declare a new variable with the `var` keyword. The scope of the declared variable is the rest of the function.

The second expression can be any WMLScript 1.1 expression that evaluates to a Boolean or an invalid value. This condition is evaluated on each pass through the loop. If the condition is true, the statement or statement block is performed. This conditional test is optional. If omitted, the condition always evaluates to true.

The third expression is generally used to update or increment the counter variable. This statement is executed as long as the condition is true.

Break Statements

The break statement terminates the current while or for loop and continues the program execution from the statement following the terminated loop. The following is an example of a break statement.

```
function testBreak(x) {
var index = 0;
    while (index < 6) {
        if (index == 3) break;
        index ++
    };
    return index*x;
};
```

NOTE: Using a break statement outside a while or a for statement generates an error.

Continue Statements

The continue statement terminates execution of a block of statements in a while or for loop and continues execution of the loop with the next iteration. The following is an example of a continue statement:

```
var index = 0
var count = 0
while (index < 5) {
    index++;
    if (index == 3)
    continue;
    count += index;
};
```

The continue statement does not terminate the execution of the loop:

▼ In a while loop, it returns to the condition

▲ In a for loop, it jumps to the update expression

NOTE: Using a continue statement outside of a while or a for statement generates an error.

Return Statements

The `return` statement is used inside the function body to specify the function's return value. The following is an example of a `return` statement:

```
function square(x) {
if (!(Lang.isFloat(x))) return invalid;
    return x * x;
}
```

If no `return` statement is specified, or none of the function `return` statements is executed, the function returns an empty string.

LIBRARIES

Libraries are named collections of functions that belong logically together. These functions can be called by using a dot (.) separator with the library name and the function name, with parameters. For example, the following calls the `elementAt` function in the String library, specifying three parameters:

```
function dummy(str) {
var i = String.elementAt(str,3,"");
};
```

Notational Conventions

Each function in the following libraries is represented by the following information, as relevant.

Function

- ▼ The function name and the number of function parameters. Function names are case sensitive. For example: abs (*value*)
- ▲ A description of what the function accomplishes.

Parameters The function parameter types. For example, *value* = number

Return Value Specifies the type of return value. For example, string or invalid.

Exceptions Describes the possible special exceptions, error codes, and corresponding return values. Standard errors, common to all functions, are not described here. For example, if value1 <= 0 and value2 < 0 is not an integer, invalid is returned.

Examples Gives a few examples of how the function could be used.
```
var a = -3;
var b = Lang.abs(a); // b = 3
```

Lang Library

The Lang library contains a set of functions that are closely related to the WMLScript 1.1 language core.

abort
`lang.abort(errorDescription)`

Aborts the interpretation of the WMLScript bytecode and returns the control to the caller of the WMLScript interpreter with the `value of errorDescription`. This function can be used to perform an abnormal exit in cases where the execution of the WMLScript should be discontinued because of serious errors detected by the calling function. If the type of the `errorDescription` is invalid, the string "invalid" is used instead.

Parameter   `errorDescription` = string

Return Value   None (this function aborts the interpretation).

Example
`Lang.abort("Error: " + errVal); //Error value is a string`

abs
`lang.abs(value)`

Returns the absolute value of the given `value`. If the given `value` is of type integer, an integer value is returned. If the given `value` is of type floating-point, a floating-point value is returned.

Parameter   `value` = number

Return Value   Number or invalid.

Examples
```
var a = -3;
var b = Lang.abs(a); //b = 3
```

characterSet
`lang.characterSet()`

Returns the character set supported by the WMLScript 1.1 interpreter. The return value is an integer that denotes a MIBenum value assigned by the Internet Assigned Numbers Authority (IANA) for all character sets.

Return Value   Integer.

Example
`var charset = Lang.characterSet(); // charset = 4 for latin1`

exit

`lang.exit(value)`

Ends the interpretation of WMLScript 1.1 bytecode and returns control to the caller of the WMLScript 1.1 interpreter with the given parameter `value`. This function can be used to perform a normal exit from a function in cases where the execution of the WMLScript should be discontinued.

Parameter   `value` = any

Return Value   None (this function ends the interpretation).

Examples
```
Lang.exit("Value: " + myVal); // Returns a string to the interpreter
Lang.exit(invalid); // Returns invalid to the interpreter
```

float

`lang.float()`

Returns true if floating-points values are supported by the interpreter being used, and false if not. This will vary according to the micro-browser being used.

Return Value   Boolean.

Example
```
var floatsSupported = Lang.float();
```

isFloat

`lang.isFloat(value)`

Returns a Boolean value that is true if the given `value` can be converted into a floating-point number using `parseFloat(value)`. Otherwise false is returned.

Parameter   `value` = any

Return Value   Boolean or invalid.

Exceptions   If the system does not support floating-point operations, an invalid value is returned.

Examples
```
var a = Lang.isFloat (" -123"); // true
var b = Lang.isFloat (" 123.33"); // true
var c = Lang.isFloat ("string"); // false
var d = Lang.isFloat ("#123.33"); // false
var e = Lang.isFloat (invalid); // invalid
```

isInt
`lang.isInt(value)`

Returns a Boolean value that is true if the given `value` can be converted into an integer by using `parseInt(value)`. Otherwise false is returned.

Parameter `value` = any

Return Value Boolean or invalid.

Examples
```
var a = Lang.isInt(" -123"); // true
var b = Lang.isInt(" 123.33"); // true
var c = Lang.isInt("string"); // false
var d = Lang.isInt("#123"); // false
var e = Lang.isInt(invalid); // invalid
```

max
`lang.max(value1, value2);`

Returns the maximum value of the given two numbers, `value1` and `value2`. The value and type returned is the same as the value and type of the selected number. The selection is done in the following way:

- ▼ Compare the numbers to select the larger one.
- ▲ If the values are equal, the first value is selected.

WMLScript 1.1 operator data type conversion rules for integers and floating-points are used to specify the data type (integer or floating-point) for comparison.

Parameters `value1` = number
              `value2` = number

Return Value Number or invalid.

Examples
```
var a = -3;
var b = Lang.abs(a);
var c = Lang.max(a, b); // c = -3
var d = Lang.max(45.5, 76); // d = 76(integer)
var e = Lang.max(45.0, 45); // e = 45.0
```

maxInt
`lang.maxInt()`

Returns the maximum integer value supported by the micro-browser on the device being used.

Return Value   Integer.

Example
```
var a = Lang.maxInt();
```

min
```
lang.min(value1, value2)
```

Returns the minimum value of the given two numbers. The value and type returned is the same as the value and type of the selected number. The selection is done in the following way:

- ▼ Compare the numbers to select the smaller one.
- ▲ If the values are equal, the first value is selected.

WMLScript operator data type conversion rules for integers and floating-points are used to specify the data type (integer or floating-point) for comparison.

Parameters   *value1* = number
              value2 = number

Return Value   Number or invalid.

Examples
```
var a = -3;
var b = Lang.abs(a);
var c = Lang.min(a, b);    // c = -3
var d = Lang.min(45, 76.3); // d = 45 (integer)
var e = Lang.min(45, 45.0); // e = 45 (integer)
```

minInt
```
lang.minInt()
```

Returns the minimum integer value supported by the micro-browser on the device being used.

Return Value   Integer.

Example
```
var a = Lang.minInt();
```

parseFloat
```
lang.parseFloat(value)
```

Returns a floating-point value defined by the string `value`. The legal floating-point syntax is specified by the WMLScript 1.1 numeric string grammar for decimal floating-point literals with the following additional parsing rule:

▼ Parsing ends when the first character is encountered that cannot be parsed as being part of the floating-point representation.

The result is the parsed string converted to a floating-point value.

Parameter   `value` = string

Return Value   Floating-point or invalid.

Exceptions   In case of a parsing error, an invalid value is returned. If the system does not support floating-point operations, an invalid value is returned.

Examples
```
var a = Lang.parseFloat("123.4");        // a = 123.4
var b = Lang.parseFloat(" +7.34e2 Hz")   // b = 7.34e2
var c = Lang.parseFloat(" 70e-2 F");     // c = 70.0e-2
var d = Lang.parseFloat("-.1 C");        // d = -0.1
var e = Lang.parseFloat(" 100 ");        // e = 100.0
var f = Lang.parseFloat("Number: 5.5");  // f = invalid
var g = Lang.parseFloat("7.3e meters");  // g = invalid
var h = Lang.parseFloat("7.3e- m/s");    // h = invalid
```

parseInt

`lang.parseInt(value)`

Returns an integer value defined by the string `value`. The legal integer syntax is specified by the WMLScript 1.1 numeric string grammar for decimal integer literals with the following additional parsing rule:

▼ Parsing ends when the first character is encountered that is not a leading "+" or "-" or decimal digit.

The result is the parsed string converted to an integer value.

Parameter   `value` = string

Return Value   Integer or invalid.

Exceptions   In case of a parsing error, an invalid value is returned.

Examples
```
var i = Lang.parseInt ("1234");       // i = 1234
var j = Lang.parseInt (" 100 m/s");   // j = 100
var k = Lang.parseInt("The larch")    // k = invalid
```

random

`lang.random(value)`

Returns an integer value with a positive value greater than or equal to 0 but less than or equal to the given `value`. The return value is chosen randomly or pseudo-randomly with approximately uniform distribution over that range, using an implementation-dependent algorithm or strategy.

If the `value` is of type floating-point, `Float.int()` is first used to calculate the actual integer value.

Parameter   `value` = integer

Return Value   Integer or invalid.

Exceptions   If `value` is equal to 0, the function returns 0. If `value` is less than 0, the function returns invalid.

Examples
```
var a = 10;
var b = Lang.random(5.1)*a;    // b = 0..50
var c = Lang.random("string"); // c = invalid
```

seed

`lang.seed(value)`

Initializes the pseudo-random number sequence and returns an empty string. If the `value` is 0 or a positive integer, the given `value` is used for initialization; otherwise a random, system-dependent initialization value is used.

If the `value` is of type floating-point, `Float.int()` is first used to calculate the actual integer *value*.

Parameter   `value` = integer

Return Value   String or invalid.

Examples
```
var a = Lang.seed(123);   // a = ""
var b = Lang.random(20);  // b = 0..20
var c = Lang.seed("seed");
// c = invalid (random seed left unchanged)
```

Float Library

The Float library contains a set of typical floating-point functions that are frequently used by applications. The implementation of these library functions is optional and implemented only by devices that can support floating-point operations. If floating-point operations are not supported, all functions in this library must return invalid.

ceil

`float.ceil(value)`

Returns the smallest integer value that is not less than the given value. If the `value` is already an integer, the result is the `value` itself.

Parameter   `value` = number

Return Value   Integer or invalid.

Examples
```
var a = 3.14;
var b = Float.ceil(a);     // b = 4
var c = Float.ceil(-2.8);  // c = -2
```

floor

`float.floor(value)`

Returns the greatest integer value that is not greater than the given value. If the `value` is already an integer, the result is the `value` itself.

Parameter   `value` = number

Return Value   Integer or invalid.

Examples
```
var a = 3.14;
var b = Float.floor(a);     // b = 3
var c = Float.floor(-2.8);  // c = -3
```

int

`float.int(value)`

Returns the integer part of the given value. If the value is already an integer, the result is the `value` itself.

Parameter   `value` = number

Return Value   Integer or invalid.

Examples
```
var a = 3.14;
var b = Float.int(a);     // b = 3
var c = Float.int(-2.8);  // c = -2
```

maxFloat

```
float.maxFloat()
```

Returns the maximum floating-point value supported by single-precision floating-point format that can be represented by the micro-browser on the device being used.

Return Value   Floating-point number - typically 3.40282347E+38.

Example
```
var a = Float.maxFloat();
```

minFloat

```
float.minFloat()
```

Returns the smallest nonzero floating-point value supported by the single-precision floating-point format that can be represented by the micro-browser on the device being used.

Return Value   Floating-point number - typically 1.17549435E–38.

Example
```
var a = Float.minFloat();
```

pow

```
float.pow(value1, value2)
```

Returns an implementation-dependent approximation of the result of raising `value1` to the power of `value2`. If `value1` is a negative number, then `value2` must be an integer.

Parameters   `value1` = number
                  `value2` = number

Return Value   Floating-point or invalid.

Exceptions   If `value1` == 0 and `value2` < 0, then invalid is returned. If `value1` < 0 and `value2` is not an integer, then invalid is returned.

Examples
```
var a = 3;
var b = Float.pow(a, 2);   // b = 9
var c = 2.78
var d = float.pow(c, 3)    // d = 2.78
```

round

```
float.round(value)
```

Returns the number value that is closest to the given *value* and is equal to a mathematical integer. If two integer number values are equally close to the *value*, the result is the largest number value. If the *value* is already an integer, the result is the *value* itself.

Parameter   *value* = number

Return Value   Integer or invalid.

Examples
```
var a = Float.round(3.5);  // a = 4
var b = Float.round(-3.5); // b = -3
var c = Float.round(0.5);  // c = 1
var d = Float.round(-0.5); // d = 0
```

sqrt
```
float.sqrt(value)
```

Returns an implementation-dependent approximation to the square root of the given *value*.

Parameter   *value* = floating-point

Return Value   Floating-point or invalid.

Exceptions   If *value* is a negative number, invalid is returned.

Examples
```
var a = 4;
var b = Float.sqrt(a); // b = 2.0
var c = Float.sqrt(5); // c = 2.2360679775
```

String Library

The String library contains a set of string functions. A string is an array of characters; each of the characters has an index with the first character in a string having an index of 0. The length of the string is the number of characters in the array.

The user of the String library can specify a special *separator* by which *elements* in a string can be separated. These elements can be accessed by specifying the separator and the element index (the first element in a string has an index 0). Each occurrence of the separator in the string separates two elements, and no escaping of separators is allowed.

A *whitespace* character is one of the following characters:

- **TAB**   Horizontal tabulation
- **VT**   Vertical tabulation
- **FF**   Form feed

- SP Space
- LF Line feed
- CR Carriage return

charAt

`string.charAt(string, index)`

Returns a new string of length 1 containing the character at the specified *index* of the given *string*. If the *index* is of type floating-point, `Float.int()` is first used to calculate the actual integer *index*.

Parameters *string* = string
index = number (the index of the character to be returned)

Return Value String or invalid.

Exceptions If *index* is out of range, an empty string ("") is returned.

Examples
```
var a = "Monday, May 24";
var b = String.charAt(a, 0);       // b = "M"
var c = String.charAt(a, 100);     // c = ""
var d = String.charAt(34, 0);      // d = "3"
var e = String.charAt(a, "first"); // e = invalid
```

compare

`string.compare(string1, string2)`

The return value indicates the lexicographic relation of *string1* to *string2*. The relation is based on the relation of the character codes in the native character set. The return value is –1 if *string1* is less than *string2*, 0 if *string1* is identical to *string2*, or 1 if *string1* is greater than *string2*.

Parameters *string1* = string
string2 = string

Return Value Integer or invalid.

Examples
```
var a = "Hello";
var b = "Hello";
var c = String.compare(a, b);          // c = 0
var d = String.compare("Bye", "Jon");  // d = -1
var e = String.compare("Jon", "Bye");  // e = 1
```

elementAt
`string.elementAt(string, index, separator)`

Searches *string* for the element enumerated by *index*, elements being separated by *separator*, and returns the corresponding element. If the *index* is less than 0, the first element is returned. If the *index* is larger than the number of elements, the last element is returned. If the *string* is an empty string, an empty string is returned.

If the *index* is of type floating-point, `Float.int()` is first used to calculate the actual *index* value.

Parameters
string = string
index = number (the index of the element to be returned)
separator = string (the first character of the string used as separator)

Return Value String or invalid.

Exceptions Returns invalid if the *separator* is an empty string ("").

Examples
```
var a = "My name is Joe; Age 50;";
var b = String.elementAt(a, 0, " ");   // b = "My"
var c = String.elementAt(a, 14, ";");  // c = ""
var d = String.elementAt(a, 1, ";");   // d = " Age 50"
```

elements
`string.elements(string, separator)`

Returns the number of elements in the given *string*, separated by the given *separator*. The empty string ("") is a valid element (thus, this function can never return a value that is less than or equal to 0).

Parameters
string = string
separator = string (the first character of the string used as separator)

Return Value Integer or invalid.

Exceptions Returns invalid if the *separator* is an empty string ("").

Examples
```
var a = "My name is Joe; Age 50;";
var b = String.elements(a, " ");    // b = 6
var c = String.elements(a, ";");    // c = 3
var d = String.elements(""; ";");   // d = 1
var e = String.elements("a", ";");  // e = 1
var f = String.elements(";", ";");  // f = 2
```

```
var g = String.elements(";;,;", ";,");
// g = 4 separator = ;
```

find

```
string.find(string, subString)
```

Returns the index of the first character in *string* that matches the requested *subString*. If no match is found, the integer value –1 is returned.

Two strings are defined to match when they are *identical*. Characters with multiple possible representations match only if they have the same representation in both strings. No case folding is performed.

Parameters   *string* = string
               subString = string

Return Value   Integer or invalid.

Examples
```
var a = "abcde";
var b = string.find(a, "cd");     // b = 2
var c = string.find(34.2, "de");  // c = -1
var d = string.find(a, "qz");     // d = -1
var e = string.find(34, "3");     // e = 0
```

format

```
string.format(format, value)
```

Converts the given *value* to a string by using the given formatting provided as a *format* string. The *format* string can contain only one format specifier, which can be located anywhere inside the string. If more than one is specified, only the first one (the leftmost) is used, and an empty string replaces the remaining specifiers. The format specifier has the following form:

```
% [width] [.precision] type
```

The *width* argument is a non-negative decimal integer controlling the minimum number of characters printed. If the number of characters in the output value is less than the specified width, blanks are added to the left until the minimum width is reached. The *width* argument never causes the *value* to be truncated. If the number of characters in the output value is greater than the specified width or, if *width* is not given, all characters of the *value* are printed (subject to the *precision* argument).

The *precision* argument specifies a non-negative decimal integer, preceded by a period (.), which can be used to set the precision of the output value. The interpretation of this value depends on the given *type*:

▼ d   Specifies the minimum number of digits to be printed. If the number of digits in the *value* is less than *precision*, the output value is padded on the left with zeroes. The *value* is not truncated when the number of digits exceeds *precision*. The default precision is 1; if *precision* is specified as 0, and the value to be converted is 0, the result is an empty string ("").

■ f   Specifies the number of digits after the decimal point. If a decimal point appears, at least one digit appears before it, and the value is rounded to the appropriate number of digits. Default precision is 6; if precision is 0 or if the period appears without a number following it, no decimal point is printed. When the number of digits after the decimal point in the *value* is less than the *precision*, the number 0 is used to fill in the columns for the needed *precision*. For example, the result of string.format("%2.3f", 1.2) will be " 1.200".

▲ s   Specifies the maximum number of characters to be printed. By default, all characters are printed. When the *width* is larger than the *precision*, the *width* is ignored.

Unlike the *width* argument, the *precision* argument can cause either truncation of the output value or rounding of a floating-point value.

The *type* argument is the only required format argument; it appears after any optional format fields. The *type* character determines whether the given *value* is interpreted as integer, floating-point, or string. The supported type arguments are the following:

▼ d   Integer; the output value has the form [-] *dddd*, where *dddd* is one or more decimal digits.

■ f   Floating-point; the output value has the form [-] *dddd.dddd*, where *dddd* is one or more decimal digits. The number of digits before the decimal point depends on the magnitude of the number, and the number of digits after the decimal point depends on the requested precision.

▲ s   String; characters are printed up to the end of the string or until the precision value is reached.

The percent character (%) in the format string can be presented by preceding it with another percent character (%%).

Parameters   *format* = string
value = any

Return Value   String or invalid.

Exceptions   Illegal format specifier results in an invalid return value.

Examples
```
var a = 45;
var b = -45;
```

```
var c = "now";
var d = 1.2345678
var e = String.format("e: %6d", a);     // e = "e: 45"
var f = String.format("%6d", b);        // f = "   -45"
var g = String.format("%6.4d", a);      // g = "  0045"
var h = String.format("%6.4d", b);      // h = " -0045"
var i = String.format("Do it %s", c);   // i = "Do it now"
var j = String.format("%3f", d);        // j = "1.234567"
var k = String.format("%10.2f%%", d);   // k = "      1.23%"
var l = String.format("%3f %2f.", d);   // l = "1.234567."
var m = String.format("%.0d", 0);       // m = ""
var n = String.format("%7d", "Int");    // n = invalid
var o = String.format("%s", true);      // o = "true".
```

insertAt

`string.insertAt(string, element, index, separator)`

Returns a string with the *element* and the corresponding *separator* (if needed) inserted at the specified element *index* of the original *string*. If the *index* is less than 0, then 0 is used as the *index*. If the *index* is larger than the number of elements, then the element is appended at the end of the *string*. If the *string* is empty, the function returns a new string with the given *element*.

If the *index* is of type floating-point, `Float.int()` is first used to calculate the actual *index* value.

Parameters *string* = string (original string)
element = string (element to be inserted)
index = number (the index of the element to be added)
separator = string (the first character of the string used as separator)

Return Value String or invalid.

Exceptions Returns invalid if the *separator* is an empty string ("").

Examples
```
var a = "B C; E";
var s = " ";
var b = String.insertAt(a, "A", 0, s);    // b = "A B C; E"
var c = String.insertAt(a, "X", 3, s);    // c = "B C; E X"
var d = String.insertAt(a, "D", 1, ";");  // d = "B C;D; E"
var e = String.insertAt(a, "F", 5, ";");  // e = "B C; E;F"
```

isEmpty

`string.isEmpty(string)`

Returns a Boolean true if the string length is 0; otherwise returns a Boolean false.

Parameter   *string* = string

Return Value   Boolean or invalid.

Examples
```
var a = "Hello";
var b = "";
var c = String.isEmpty(a);     // c = false
var d = String.isEmpty(b);     // d = true
var e = String.isEmpty(true);  // e = false
```

length
`string.length(string)`

Returns the length (number of characters) of the given *string*.

Parameter   *string* = string

Return Value   Integer or invalid.

Examples
```
var a = "ABC";
var b = String.length(a);    // b = 3
var c = String.length("")    // c = 0
var d = String.length(342);  // d = 3
```

removeAt
`string.removeAt(string, index, separator)`

Returns a new string where the element and the corresponding *separator* (if existing) with the given *index* are removed from the given *string*. If the *index* is less than 0, then the first element is removed. If the *index* is larger than the number of elements, the last element is removed. If the *string* is empty, the function returns a new empty string.

If the *index* is of type floating-point, `Float.int()` is first used to calculate the actual *index* value.

Parameters   *string* = string
index = number (the index of the element to be deleted)
separator = string (the first character of the string used as separator)

Return Value   String or invalid.

Exceptions   Returns invalid if the *separator* is an empty string ("").

Examples

```
var a = "A A; B C D";
var s = "";
var b = String.removeAt(a, 1, s);    // b = "A B C D"
var c = String.removeAt(a, 0, ";");  // c = " B C D"
var d = String.removeAt(a, 14, ";"); // d = "A A"
```

replace

`string.replace(string, oldSubString, newSubString)`

Returns a new string resulting from replacing all occurrences of *oldSubString* in this *string* with *newSubString*.

Two strings are defined to match when they are *identical*. Characters with multiple possible representations match only if they have the same representation in both strings. No case folding is performed.

Parameters   *string* = string
   oldSubString = string
   newSubString = string

Return Value   String or invalid.

Examples

```
var a = "Hello Christina. What is up Christina?";
var newName = "Marie"
var oldName = "Christina"
var c = String.replace(a, oldName, newName);
// c = "Hello Marie. What is up Marie?"
var d = String.replace(a, newName, oldName);
// d = "Hello Christina. What is up Christina?"
```

replaceAt

`string.replaceAt(string, element, index, separator)`

Returns a string with the current element at the specified *index* replaced with the given *element*. If the *index* is less than 0, the first element is replaced. If the *index* is larger than the number of elements, the last element is replaced. If the *string* is empty, the function returns a new string with the given *element*.

If the *index* is of type floating-point, `Float.int()` is first used to calculate the actual *index* value.

Parameters   *string* = string
   element = string
   index = number (the index of the element to be replaced)
   separator = string (the first character of the string used as separator)

Return Value   String or invalid.

Exceptions   Returns invalid if the *separator* is an empty string ("").

Examples
```
var a = "B C; E";
var s = "";
var b = String.replaceAt(a, "A", 0, s);    // b = "A C; E"
var c = String.replaceAt(a, "F", 5, ";");  // c = "B C;F"
```

squeeze

```
string.squeeze(string)
```

Returns a string where all consecutive series of white spaces within the *string* are reduced to one.

Parameter   *string* = string

Return Value   String or invalid.

Examples
```
var a = "Hello";
var b = "   Bye  Jon  .  See you!   ";
var c = String.squeeze(a); // c = "Hello";
var d = String.squeeze(b)  // d = " Bye Jon . See you! "
```

subString

```
string.subString(string, startIndex, length)
```

Returns a new string that is a substring of the given *string*. The substring begins at the specified *startIndex* and its length (number of characters) is the given *length*. If the *startIndex* is less than 0, then 0 is used for the *startIndex*. If the *length* is larger than the remaining number of characters in the *string*, the *length* is replaced with the number of remaining characters.

If the *startIndex* or the *length* is of type floating-point, Float.int() is first used to calculate the actual integer value.

Parameters   *string* = string
startIndex = number (the beginning index, inclusive)
length = number (the length of the substring)

Return Value   String or invalid.

Exceptions   If *startIndex* is larger than the last index, an empty string ("") is returned. If *length* <= 0, an empty string ("") is returned.

Examples

```
var a = "ABCD";
var b = String.subString(a, 1, 2);      // b = "BC"
var c = String.subString(a, 2, 5);      // c = "CD"
var d = String.subString(1234, 0, 2);   // d = "12"
```

toString

`string.toString(value)`

Returns a string representation of the given *value*. This function performs exactly the same conversions as supported by WMLScript 1.1 (automatic conversion from Boolean, integer, and floating-point values to strings) except that an invalid *value* returns the string "invalid".

Parameter   *value* = any

Return Value   String.

Examples

```
var a = String.toString(12);    // a = "12"
var b = String.toString(true)   // b = "true"
```

trim

`string.trim(string)`

Returns a string where all trailing and leading white spaces in the given *string* have been trimmed.

Parameter   *string* = string

Return Value   String or invalid.

Examples

```
var a = "Hello   ";
var b = " Bye Jon . See you!   ";
var c = String.trim(a);   // c = "Hello";
var d = String.trim(b)    // d = "Bye Jon . See you!"
```

URL Library

The URL library contains a set of functions for handling absolute URLs and relative URLs. The general URL syntax is:

`<scheme>://<host>:<port>/<path>;<params>?<query>#<fragment>`

escapeString
`url.escapeString(string)`

This function computes a new version of a *string* value in which special characters have been replaced by a hexadecimal escape sequence; a two-digit escape sequence of the form `%xx` is used.

The characters to be escaped are the following:

- ▼ **Control characters**   <US-ASCII coded characters 00-1F and 7F>
- ■ **Space**   <US-ASCII coded character 20 hexadecimal>
- ■ **Reserved**   ; / ? : @ & = + $,
- ■ **Unwise**   ( ) | \ ^ [ ] `
- ▲ **Delims**   < > # % "

The given string is escaped just as it is given. No URL parsing is performed.

Parameter   *string* = string

Return Value   String or invalid.

Exceptions   If *string* contains characters that are not part of the US-ASCII character set, an invalid value is returned.

Example
```
var a = URL.escapeString("http://w.h.com/dck?x=\u007f#crd");
// a = "http%3a%2f%2fw.h.com%2fdck%3fx%3d%7f%23crd"
```

getBase
`url.getBase()`

Returns an absolute URL (without the fragment) of the current WMLScript 1.1 file.

Return Value   String.

Example
```
var a = URL.getBase();
// a = http://www.host.com/test.scr
```

getFragment
`url.getFragment(url)`

Returns the fragment used in the given *url*. If no fragment is specified, an empty string is returned. Both absolute and relative URLs are supported. Relative URLs are not resolved into absolute URLs.

Parameter   `url` = string

Return Value   String or invalid.

Exceptions   If an invalid URL syntax is encountered while extracting the fragment, an invalid value is returned.

Example
```
var a = URL.getFragment("http://w.h.com/cont#frag");  // a = "frag"
```

getHost
```
url.getHost(url)
```

Returns the host specified in the given `url`. Both absolute and relative URLs are supported. Relative URLs are not resolved into absolute URLs.

Parameter   `url` = string

Return Value   String or invalid.

Exceptions   If an invalid URL syntax is encountered while extracting the host part, an invalid value is returned.

Examples
```
var a = URL.getHost("http://w.h.com/path#frag");  // a = "w.h.com"
var b = URL.getHost("path#frag");                  // b = ""
```

getParameters
```
url.getParameters(url)
```

Returns the parameters used in the given `url`. If no parameters are specified, an empty string is returned. Both absolute and relative URLs are supported. Relative URLs are not resolved into absolute URLs.

Parameter   `url` = string

Return Value   String or invalid.

Exceptions   If an invalid URL syntax is encountered while extracting the parameters, an invalid value is returned.

Examples
```
a = URL.getParameters("http://w.h.com/script;3;27x=1&y=3");
// a = "3;2"
b = URL.getParameters("../script;3;2?x=1&y=3");
// b = "3;2"
```

getPath
url.getPath(*url*)

Returns the path specified in the given *url*. Both absolute and relative URLs are supported. Relative URLs are not resolved into absolute URLs.

Parameter   *url* = string

Return Value   String or invalid.

Exceptions   If an invalid URL syntax is encountered while extracting the path, an invalid value is returned.

Examples
```
a = URL.getPath("http://w.h.com/home/sub/comp#frag");
// a = "/home/sub/comp"
b = URL.getPath("../home/sub/comp#frag");
// b = "../home/sub/comp"
```

getPort
url.getPort(*url*)

Returns the port number specified in the given *url*. If no port is specified, an empty string is returned. Both absolute and relative URLs are supported. Relative URLs are not resolved into absolute URLs.

Parameter   *url* = string

Return Value   String or invalid.

Exceptions   If an invalid URL syntax is encountered while extracting the port number, an invalid value is returned.

Examples
```
var a = URL.getPort("http://w.h.com:80/path#frag");  // a = "80"
var b = URL.getPort("http://w.h.com/path#frag");     // b = ""
```

getQuery
url.getQuery(*url*)

Returns the query part specified in the given *url*. If no query part is specified, an empty string is returned. Both absolute and relative URLs are supported. Relative URLs are not resolved into absolute URLs.

Parameter   *url* = string

Return Value   String or invalid.

Exceptions   If an invalid URL syntax is encountered while extracting the query part, an invalid value is returned.

Examples
```
a = URL.getParameters("http://w.h.com/home;3;2?x=1&y=3");
// a = "x=1&y=3"
var base = URL.getBase();
// base = http://www.host.com/current.scr
```

getReferer
`url.getReferer()`

Returns the smallest relative URL (relative to the base URL of the current file) to the resource that called the current file. Local function calls do not change the referer. If the current file does not have a referer, an empty string ("") is returned.

Return Value   String.

Example
```
var referer = URL.getReferer();
// referer = "app.wml"
```

getScheme
`url.getScheme(url)`

Returns the scheme used in the given `url`. Both absolute and relative URLs are supported. Relative URLs are not resolved into absolute URLs.

Parameter   `url` = string

Return Value   String or invalid.

Exceptions   If an invalid URL syntax is encountered while extracting the scheme, an invalid value is returned.

Examples
```
var a = URL.getScheme("http://w.h.com/path#frag");
// a = "http"
var b = URL.getScheme("w.h.com/path#frag");
// b = ""
```

isValid
`url.isValid(url)`

Returns true if the given `url` has the right URL syntax, otherwise returns false. Both absolute and relative URLs are supported. Relative URLs are not resolved into absolute URLs.

Parameter   `url` = string

Return Value   Boolean or invalid.

Examples
```
var a = URL.isValid("http://w.hst.com/script#func()");
// a = true
var b = URL.isValid("../common#test()");
// b = true
var c = URL.isValid("experimental?://www.host.com/cont");
// c = false
```

loadString
`url.loadString(url, contentType)`

Returns the content denoted by the given absolute `url` and the `contentType`. The given `contentType` is erroneous if it does not comply with the following rules:

- ▼ Only one content type can be specified. The whole string must match with only one content type and no extra leading or trailing spaces are allowed.
- ▲ The type must be text but the subtype can be anything. Thus, the type prefix must be "text/".

The behavior of this function is the following:

1. The content with the given `contentType` and `url` is loaded. The rest of the attributes needed for the content load are specified by the default settings of the user agent.
2. If the load is successful and the returned content type matches the given `contentType`, the content is converted to a string and returned.
3. If the load is unsuccessful or the returned content is of the wrong content type, a scheme-specific error code is returned.

Parameters   `url` = string
  `contentType` = string

Return Value   String, integer, or invalid.

Exceptions   Returns an integer error code that depends on the used URL scheme in case the load fails. If HTTP or WSP schemes are used, HTTP error codes are returned.
If an erroneous `contentType` is given, an invalid value is returned.

Examples
```
var myUrl = "http://www.host.com/vcards/myaddr.vcf";
myCard = URL.loadString(myUrl, "text/x-vcard");
```

resolve
`url.resolve(baseUrl, embeddedUrl)`

Returns an absolute URL from the given *baseUrl* and the *embeddedUrl*. If the *embeddedUrl* is already an absolute URL, the function returns it without modification.

Parameters *baseUrl* = string
 embeddedUrl = string

Return Value String or invalid.

Exceptions If an invalid URL syntax is encountered as part of the resolution, an invalid value is returned.

Example
```
var a = URL.resolve("http://foo.com/","foo.vcf");
// a = http://foo.com/foo.vcf
```

unescapeString
`url.unescapeString(string)`

The unescape function computes a new version of a *string* value in which each escape sequence of the sort that might be introduced by the URL.escapeString() function is replaced with the character that it represents. The given *string* is unescaped just as it was passed. No URL parsing is performed.

Parameter *string* = string

Return Value String or invalid.

Exceptions If *string* contains characters that are not part of the US-ASCII character set, an invalid value is returned.

Examples
```
var a = "http%3a%2f%2fw.h.com%2fdck%3fx%3d12%23crd";
var b = URL.unescapeString(a);
// b = http://w.h.com/dck?x=12#crd
```

WMLBrowser Library

The WMLBrowser library contains functions by which WMLScript 1.1 can access the associated WML context. These functions must not have any side effects, and they must return invalid in the following cases:

- ▼ If the system does not support the WML browser.
- ▲ If the WMLScript interpreter is not invoked by the WML browser.

getCurrentCard
`WMLBrowser.getCurrentCard()`

Returns the smallest relative URL (relative to the base of the current file) specifying the card (if any) currently being processed by the WML browser.

The function returns an absolute URL if the WML deck containing the current card does not have the same base as the current file.

Return Value String or invalid.

Exceptions Returns invalid if there is no current card.

Example
```
var a = WMLBrowser.getCurrentCard();
// a = "deck#input"
```

getVar
`WMLBrowser.getVar(name)`

Returns the value of the variable with the given *name* in the current browser context. Returns an empty string ("") if the given variable does not exist. The variable name must follow the syntax specified by WML.

Parameter *name* = string

Return Value String or invalid.

Exceptions If the syntax of the variable name is incorrect, an invalid value is returned.

Example
```
var a = WMLBrowser.getVar("name");
// a = "Jon" or whatever value the variable has.
```

go
`WMLBrowser.go(url)`

Specifies the content denoted by the given *url* to be loaded. This function has the same semantics as the go task in WML. The content is loaded only after the WML browser receives control from the WMLScript interpreter (after the WMLScript invocation is finished). No content is loaded if the given *url* is an empty string ("").

The `go()` and `prev()` library functions override each other. Both of these library functions can be called multiple times before returning control to the WML browser.

However, only the settings of the last call stay in effect. In particular, if the last call to `go()` or `prev()` sets the URL to an empty string (""), all requests are effectively cancelled.

This function returns an empty string ("").

Parameter   `url` = string

Return Value   An empty string ("").

Example
```
var card = "http://www.host.com/loc/app.dck#start";
WMLBrowser.go(card);
```

newContext
`WMLBrowser.newContext()`

Clears the current WML browser context and returns an empty string (""). This function has the same semantics as the `newcontext` attribute in WML.

Return Value   An empty string ("").

Example
```
WMLBrowser.newContext();
```

prev
`WMLBrowser.prev()`

Signals the WML browser to go back to the previous WML card. This function has the same semantics as the `prev` task in WML. The previous card is loaded only after the WML browser receives control from the WMLScript 1.1 interpreter (after the WMLScript 1.1 invocation is finished).

The `prev()` and `go()` library functions override each other. Both of these library functions can be called multiple times before returning control back to the WML browser. However, only the settings of the last call stay in effect. In particular, if the last call to `go()` or `prev()` set the URL to an empty string (""), all requests are effectively cancelled.

Return Value   An empty string ("").

Example
```
WMLBrowser.prev();
```

refresh
`WMLBrowser.refresh()`

Forces the WML browser to update its context and returns an empty string. This function has the same semantics as the `refresh` task in WML.

Return Value An empty string ("").

Exceptions Returns invalid if there is no current card.

Example
```
function convert2Peso(){
var dollars = WMLBrowser.getVar("amount");
var dol2peso = 10.2;
var newAmt = dollars*dol2peso;
WMLBrowser.setVar("amount", newAmt);
WMLBrowser.refresh();
}
```

setVar
`WMLBrowser.setVar(name, value)`

Returns true if the variable with the given *name* is successfully set to contain the given *value* in the current browser context; false otherwise. The variable name and its value must follow the syntax specified by WML. The variable value must be legal XML CDATA.

Parameters *name* = string
value = string

Return Value Boolean or invalid.

Exceptions If the syntax of the variable name or its value is incorrect, an invalid value is returned.

Example
```
var a = WMLBrowser.setVar("name", "Mary"); // a = true
```

Dialogs Library
The Dialogs library contains a set of typical user-interface functions.

alert
`dialogs.alert(message)`

Displays the given *message* to the user, waits for the user confirmation, and returns an empty string ("").

Parameter *message* = string

Return Value   String or invalid.

Example
```
function testValue(textElement) {
if (String.length(textElement) > 8) {
    Dialogs.alert("Enter name < 8 chars!");
   };
};
```

confirm

`dialogs.confirm(message, ok, cancel)`

Displays the given *message* and two reply alternatives: *ok* and *cancel*. Waits for the user to select one of the reply alternatives and returns true for *ok* and false for *cancel*.

Parameters
   message = string
   ok = string (text; empty string results in the default implementation-dependent text)
   cancel = string (text; empty string results in the default implementation-dependent text)

Return Value   Boolean or invalid.

Example
```
function onAbort() {
return Dialogs.confirm("Are you sure?","Well...","Yes");
};
```

prompt

`dialogs.prompt(message, defaultInput)`

Displays the given *message* and prompts for user input. The *defaultInput* parameter contains the initial content for the user input. Returns the user input.

Parameters   message = string
            defaultInput = string

Return Value   String or invalid.

Example
```
var a = "234-1234";
var b = Dialogs.prompt("Phone number: ",a);
```

Console Library

The Dialogs library contains a set of typical user-interface functions.

print
`Console.print(string)`

Used for debugging. Converts any value to a string and prints it to the SDK Phone Information window.

printLn
`Console.println(string)`

Used for debugging. Converts any value to a string and prints it to the SDK Phone Information window. This call is the same as `console.print` except that it adds a newline character to the end of the string.

GLOSSARY

3G   Third generation. Radio link protocols such as W-CDMA, CDMA2000, and TDMA offer higher data rates than second generation digital.

ACK   Acknowledgment. Indication of successful completion of task.

ALOHA   A channel access method whereby a message is sent in a non-scheduled fashion. If a collision occurs, a back-off occurs and another attempt is made.

AM   Amplitude Modulation. A modulation technique that involves changes to a signal carrier's amplitude.

APDU   Application Protocol Data Unit. A packet of information.

API   Application Program Interface. A collection of defined software functions, classes, methods, etc., generally packaged as software binary libraries, that are used by programmers to take advantage of various system features.

ASCII   American Standard Code for Information Interchange.

ASN.1   Abstract Syntax Notation One. An encoding technique that ensures different data types can be exchanged between systems having different data representations caused by different byte ordering, character sets, etc.

BER   Basic Encoding Rules. Rules that determine how different data types are encoded.

BLOB   Binary Large Object. A data type used by WMtp that may be any type of binary data, originally used for compressed voice files.

bps   Bits per second.

Bytecode   Content encoding in which the content is typically a set of low-level instructions and operands for a targeted hardware (or virtual) machine.

Caller   The person or machine that originates a messaging session with the paging system.

Capcode   A pager's internal address.

CDMA2000   A third generation wireless protocol which is an outgrowth of CDMA 1.

Client   A device or application that initiates a request for connection to a server.

Code plug   The program and configuration image in a pager's memory.

Content   Data stored or generated at an origin server. Content is typically displayed or interpreted by a user agent in response to a user request.

Content encoding   When used as a verb, content encoding indicates the act of converting a data object from one format to another. Typically the resulting format requires less physical space than the original, is easier to process or store, or is encrypted. When used as a noun, content encoding specifies a particular format or encoding standard or process.

Content format   Actual representation of content.

CRC   Cyclic Redundancy Check. An error detection code.

CSU / DSU   Channel Service Unit / Digital Service Unit. Devices that encapsulate information into the proper framing before distribution over a WAN.

Device   A network entity that is capable of sending and receiving packets of information and that has a unique device address. A device can act as both a client and a server within a given context or across multiple contexts. For example, a device can service a number of clients (as a server) while being a client to another server.

DSP   Digital Signal Processor. A programmable device that is often used to provide embedded logic in wireless and other products.

DTD   Document Type Definition. The definition states which elements can be nested within others. A DTD defines the names and contents of all elements permissible in a certain document, how often an element may appear, the order in which the elements must appear, whether the start or end tag may be omitted, the contents of all elements, that is, the names of the other generic identifiers that are allowed to appear inside them, the attributes and their default values and the names of the reference symbols that may be used.

DTMF   Dual Tone Multi-Frequency. Tones generated in telephones used for in-band signaling.

DU   Data Unit. Packets of information.

Element   An element specifies the markup and structural information in a WML deck. Some elements contain a start and end tag such as the <p> and </p> tag, others are single elements such as the <br/> tag.

ERMES   European Radio Message System. A radio protocol developed by operators in Europe.

FDD   Frequency Division Duplex.

FLEX   The high speed one-way over the air paging protocol developed by Motorola.

FMS   FLEX Messaging Server. A product from Motorola that provides email and Internet access to paging systems.

Frame Relay   A simple connection oriented layer 2 protocol defined by the ITU-T used to transfer information between two end points.

FSK   Frequency Shift Keying. A modulation technique that involves changing the frequency of a signal carrier to convey digital information.

FTP   File Transfer Protocol. A protocol used to reliably send files across a network.

GOTAP   Generic Over The Air Programming. A set of generic commands used to modify specific attributes or configurations in pagers. These generic commands can be translated into pager specific commands needed to modify a pager's code plug using OTAP.

GPRS   General Packet Radio Service. A packet-based bearer that is being introduced on many GSM and TDMA mobile networks. Its main difference compared to GSM, which uses a predominantly circuit switched protocol, is that GPRS only uses the network when data is to be sent. GPRS enables users to send data at speeds of up to 115KB per second. It is often referred to as "always on" technology, and brings e-mail and Internet access capabilities to a GSM network.

GPS   Global Positioning Satellite. A system of satellites maintained by the US government that provides location information and precise timing. Used by paging systems to synchronize simulcast.

GSM   Global System for Mobile Communications. A second-generation digital cellular radio standard developed in Europe but widely adopted around the world.

HDML   Handheld Markup Language. A markup language optimized for use in wireless hand held devices.

Hex   Hexadecimal.

HTML   Hypertext Markup Language. A markup language widely used to send information throughout the Internet. Suitable for wired environments.

HTTP   Hypertext Transfer Protocol. A connectionless, stateless protocol well suited for browsing sites on the Internet.

Inbound   Message direction moving from a wireless device to the infrastructure.

InFLEXion   An over the air RF protocol used to send compressed voice messages from transmitters to pagers that support the protocol.

IP   Internet Protocol. The packet based network layer protocol used in many data networks. Predominant network layer protocol used to send data across the Internet.

IPP   Inbound Paging Protocol. An open paging protocol used to send information from paging receivers to the system controller. It uses UDP/IP.

JavaScript A de facto standard language that can be used to add dynamic behavior to HTML documents. JavaScript is one of the originating technologies of ECMAScript.

kbps Kilobits per second. A data rate measure in units of a thousand bits per second.

kHz Kilohertz. A frequency measure in units of a thousand cycles per second.

MCR Multiple Choice Response. The ability to provide a limited set of responses that a user can select from. In paging systems, the selection number rather than the text of the selection is sent across the paging system, resulting in more efficient use of the RF and network bandwidth.

MHz Megahertz. A frequency measure in units of a million cycles per second.

MIME Multipurpose Internet Mail Extensions. An RFC that devices methods of encoding various data types for use in sending them between systems on the Internet. Originally defined for mail enclosures, but extended to data types that can be sent using HTTP.

MS-H Messaging Switch – Home. Nomenclature used in WMtp to identify a node that contains the subscriber's record that a caller wishes to page.

MS-I Messaging Switch – Input. Nomenclature used in WMtp to identify a node that handles the receipt of message request, typically from the PSTN. MS-I does not contain the subscriber record that the caller wishes to page.

MS-O Messaging Switch – Output. Nomenclature used in WMtp to identify a node that handles message scheduling and encoding prior to delivery to transmitters.

NAK Negative Acknowledgment. Indication of failure in completion of task.

OAP Operator Assisted Paging. System involving human operators who receive and dispatch page requests, typically using display screens and systems that interface to a paging system.

Origin Server The server on which a given resource resides or is to be created. Often referred to as a web server or an HTTP server.

OTA or OTAP Over The Air Programming. The ability to change information in a wireless device by sending information over the radio link.

Outbound Message Direction moving toward a wireless device from the infrastructure.

PC Personal Computer.

PCS Personal Communications System. Cell based messaging systems, originally focused on enhanced voice services.

PDA   Personal Digital Assistant. Usually a hand-held device, such as the Palm Pilot. WAP isn't just for cellular phones.

PDU   Paging Data Unit. A packet of information containing paging data to be sent to a pager.

PIM   Personal Information Manager. A device that contains information organizers such as schedulers, to do lists, etc.

PIN   Personal Identification Number. A number used to identify a subscriber.

POCSAG   Post Office Code Standardization Advisory Group. The name of a earlier generation one-way paging protocol featuring fairly slow raw data rates ranging from 512 to 2400 bps.

PPP   Point-to-Point Protocol. A connection oriented, layer 2 protocol used with TCP/IP applications.

PSTN   Public Switched Telephone Network. The circuit switches telephone system.

QWERTY Keyboard   A standard typewriter keyboard. So called because these are the first six characters of the top line of letter keys.

Radio group   A group of subscribers who share a common broadcast address in their pagers. Messages sent to this common address are received by all subscribers in the group. A form of broadcast messaging.

Resource   A network data object or service that can be identified by a URL. Resources may be available in multiple representations (for example, multiple languages, data formats, size, and resolutions) or vary in other ways

RF   Radio Frequency.

RFC   Request For Comment.

RIC   Radio Identification Code. The capcode of a pager.

ROSE   Remote Operation Service Element. An inter-node message exchange technique that supports a request / response paradigm.

RS-232   A specification for serial communications.

SDK   Software Development Kit.

Server   A device (or application) that passively waits for connection requests from one or more clients. A server may accept or reject a connection request from a client.

SGML   Standardised Generalised Markup Language.

SLIP   Serial Link Interface Protocol.

SMTP   Simple Mail Transfer Protocol. A protocol used to transfer mail or other data between systems.

SNMP   Simple Network Management Protocol. A simple light weight protocol used to manage devices, typically in a computer or communications system, including the networks.

SNPP   Simple Network Paging Protocol. A paging protocol used to send messages to a paging terminal.

SSB   Single Side Band. A modulation technique which suppresses one side band of an AM signal as a method to conserve bandwidth.

Subscriber   An individual who purchases service.

Subscriber ID   The identifier of a subscriber. In paging systems this may be a PIN or an actual telephone number.

Subscriber profile   Information maintained in a paging terminal describing the services available to a subscriber.

T1 / E1   Digital circuit switches telephone links operating at 1.544 Mbps and 2.048 Mbps, respectively.

TAP   Telocator Alphanumeric Protocol. A paging protocol used to send alphanumeric messages to a paging terminal.

TCP   Transmission Control Protocol. A packet protocol used at the transport layer, which is guaranteed to be reliable. It uses IP as the network protocol.

TDD   Time Division Duplex.

Telematics   Telematics is an emerging industry that offers location based voice and data communication tools. Telematics provides smart information, tailored to where customers are and to what they're doing. This provides enhanced security, navigation, and convenience to mobile consumers.

Telnet   A terminal emulation protocol used to remotely access a computer.

Terminal group   A group of subscribers maintained as a list by a paging terminal so that sending a message to the group results in sending individual messages to each member of the group. One form of "broadcast" messaging.

TNPP   Telocator Network Paging Protocol.

UA   User agent. Software that interprets WML, WMLScript, WTAI and other forms of code.

UDP   User datagram protocol. A packet protocol used at the transport layer, which is not guaranteed to be reliable. It uses IP as the network protocol.

User   A person who interacts with a user agent to view, hear, or otherwise use rendered content

User Agent   A user agent is any WAP device, whether it is a mobile phone, a handheld device such as the Palm or HP personal digital assistants, a pager or even a household refrigerator that has been WAP enabled. Any software or device that interprets WML, WMLScript 1.1, or resources. This may include textual browsers, voice browsers, search engines, and so on.

UTC   Universal Time Coordinated. Time is seconds from January 1, 1970. The time standard used in most UNIX systems.

VoxML   Voice eXtensible Markup Language. A markup language that includes data types useful in voice applications.

W3C   World Wide Web Consortium. http://www.w3c.org/

WAE   Wireless Application Environment. The Wireless Application Environment specifies a general-purpose application environment based fundamentally on Internet technologies and philosophies. WAE specifies an environment intended for the development and execution of portable applications and services.

WAN   Wide Area Network. A network that generally involves long distance links.

WAP   Wireless Application Protocol. A specification for technology useful in developing wireless applications and services. It is actually composed of several different protocols that make up the overall protocol referred to as WAP. The layer rpotocols are: The Application layer (WAE), the Session layer (WSP), the Transaction layer (WTP), the Security layer (WTLS) and the Transport layer (WDP).

WAP Gateway   A WAP gateway is a two-way device. Looking at it from the WAP device's side, since a WAP device can only understand WML in its binary format, the function of the WAP gateway is to convert content into this format. Looking at it from the HTTP server's side, the WAP gateway can provide additional information about the WAP device through the HTTP headers, for instance the subscriber number of a WAP capable cellular phone, its cell ID and even things like location information (whenever that becomes available).

WAP Server   A WAP server by itself is really nothing more than an HTTP server - ie. a web server. Nokia has brought out a product they call a WAP server which is a WAP gateway and an HTTP server all in one. This is actually content providing servers and a WAP gateway. The gateway takes care of the gateway stuff, and the web server provides the contents.

W-CDMA   Wideband CDMA. A third-generation, wireless protocol that is a migration path for GSM.

WDP   Wireless Datagram Protocol. This is the bottom layer of the WAP "stack" of protocols, and is the firewall between the rest of the WAP layers above it and the many bearer services available.

Web server   A network host that acts as an HTTP server.

WML   Wireless Markup Language. A markup language optimized for use in wireless devices with limited capabilities.

WML Card   A single WML block of navigation and user interface in a WML deck. A WML card must exist inside a WML deck containing one or more cards. WML decks are XML documents.

WML Deck   A collection of WML cards. The whole deck is loaded when the browser requests a URL, and access to individual cards in the deck can be specified in the URL.

WMLScript   Scripting language for WAP devices. Based on JavaScript, but less powerful.

WMtp   Wireless Messaging Transfer Protocol. A packet based paging protocol developed by Glenayre that is used to exchange messages between paging terminals and controllers in a paging system.

WSP   Wireless Session Protocol. The Wireless Session Protocol provides the upper-level application layer of WAP with a consistent interface for the organized exchange of content between client/server applications.

WTLS   Wireless Transport Layer Security. This is an optional layer that provides authentication and secure connections between applications. It is the WAP equivalent to the SSL (Secure Sockets Layer) widely used in HTML.

WTP   Wireless Transaction Protocol. The Transaction layer provides different methods for performing transactions, to a varying degree of liability.

X.25   An ITU-T recommendation published in 1976 that defines a packet based interface protocol.

XML   Extensible Markup Language. W3C's standard for Internet Markup Languages. A markup language that can be extended by defining new data types. WML is one of these languages. XML is a subset of SGML.

INDEX

A

Abort function, 267
Abs function, 267
Access control, 156–157, 251–252
Action/Navigation card, 48
Active server pages (ASP), 161–167
Adding functionality with WMLScript, 117–158
ADO (ActiveX Data Objects), 168–170
    miscellaneous notes about connections, 170
    physically connecting to Database, 168
    querying databases, 169
    tidying up, 170
    using returned data, 169–170
Agents, user, 2
Alert function, 293–294
API (Application Programming Interface), 164
Application-design process, 32–37
    customers for, 32
    designing phones for, 33
    drawing screens for applications, 34
    filling in gaps, 36
    micro-browser design, 32
    problems of users, 33–34
    solving problems that are visualized, 34
    testing final products, 36–37
    testing mock-ups, 34
    testing prototypes for usability, 36
    writing prototypes, 35–36
    writing pseudo codes, 35
Application, dynamic WAP, 171–193
Application levels, 67
Application objects, 167
Application personalization, 44
Application Programming Interface (API), 164
Applications
    designing good WAP, 30–37
    drawing screens for, 34
    killer, 30
    making of good, 23–37
    for telematics, 229–230
Architecture, WAP, 8–15
Arguments, 141
Arithmetic operators, 125, 254–255
Arrays, 152–154
ASP (active server pages), 161–167
ASP and WAP, 162–164
ASP object model, 164–167
    Application objects, 167
    Request object, 165–166

307

Response object, 164–165
Server objects, 167
Session object, 166
Assignment operators, 125, 253–254
Attributes
 format, 116
 name, 111
 onpick, 90, 92, 102–103
 value, 111
Automated converters, fully, 197
Average latency, 12

B

Back button, 50–51
Bad-message response, 227
Bandwidth, low, 11–12, 42
Battery power, limited, 14–15
Bearer level, 67
Beginning of File (BOF), 180
Bitwise operators, 126
Block comments, 243
Block statements, 262–263
Blocks, code, 122
Bluetooth technology, 234, 235–237
BOF (Beginning of File), 180
Boolean variables, 246
Brackets, 134
Branding images, 77
Break statements, 138–139, 265
Breaks, line, 120, 242
Browsers
 targeting market micro, 29–30
 WAP micro, 27
Building
 databases, 173–176
 services sites, 74–75
Business case for WAP, 15–21
Button, Back, 50–51

C

Caches, using, 45–46
Calling functions, 142–145
Calls
 external function, 261
 function, 260–261
 library function, 262
 local script function, 261

Card elements, 105
 placing event types directly into, 102
Cards
 contents of, 75–76
 defined, 13
 display, 49–50
 entry, 49
 first, 72
 Help, 76
 Navigation-Only, 48
 second, 72–73
 static, 67
 static data hard-coded in, 160
 types of WML, 47–50
Cards, choice, 47–49
 Action/Navigation card, 48
 Change option, 49
 data lists, 47–48
 Navigation-Only card, 48
 Pick option, 48
Cards in WAP, pages are called, 69
Care, customer, 18–19
Case sensitivity, 120, 242
Ceil function, 273
Certificates, digital, 218
Change option, 49
Characters, variables and, 105
CharacterSet function, 267
CharAt function, 276
Choice cards, 47–49
 Action/Navigation card, 48
 Change option, 49
 data lists, 47–48
 Navigation-Only card, 48
 Pick option, 48
Chprefs.asp, 186–188
Ciphers, symmetric, 216, 217
Client-capability query, 227
Client-side scripting language, WMLScript
 is, 118
Code blocks, 122
Code, writing, 176–193
 chprefs.asp, 186–188
 Default.wml, 177–178
 login.asp, 178–181
 menu.asp, 181–184
 prefs.asp, 184–186
 results.asp, 190–193
 wrprefs.asp, 188–190
Codes
 format codes, 115

Index

template, 92, 93
writing pseudo, 35
writing WML, 68–69
Coding principles, general, 157–158
Comma operators, 257
Commas, 131–132
Comments, 120–121, 243
block, 243
line, 243
Commerce, M, 213–221
Commerce, mobile, 2
Compare function, 276
Comparison operators, 128, 129–131, 256–257
Compiler, WMLScript, 242
Concatenation, string, 131
Conditional operators, 133–134, 258
Configurable converters, 197–198
Confirm function, 294
Connections, miscellaneous notes about, 170
Console library, 295
Constants, 244–246
Boolean variables, 246
floating-point, 244–245
integer, 244
invalid variables, 246
string, 245–246
Constructs, control, 135–139
if statements, 135–137
skipping unnecessary loop statements, 138–139
for statements, 137–138
stopping loops, 138–139
while statements, 137
Consumers, benefits for, 5–6
Content limitations, 31
Continue statements, 139, 265
Control, access, 251–252
Control constructs, 135–139
if statements, 135–137
skipping unnecessary loop statements, 138–139
for statements, 137–138
stopping loops, 138–139
while statements, 137
Conventions
choice of, 141
notational, 266
Conversion, demonstration HTML, 202–210
Conversion rules, HTML to WML, 199–202
Converters
configurable, 197–198
fully automated, 197

Converting existing HTML Web sites to WAP, 196–202
Converting existing Web sites, 195–211
Converting images, 62
Corporate images, 77
CPU, limited, 14
Creating
DSNs (data source names), 175–176
links, 80–81
Cryptography, 216–218
public key, 217–218
symmetric, 217
Customer care, 18–19

D

Dailogs library, 293–294
Data entry field, password, 178
Data formatting, 112–116
emptyok, 113
format, 114–116
maxlength, 112
size, 112
type, 112–113
Data hard-coded in cards, static, 160
Data; *See also* Metadata
field entry, 45
flow, 172–173
lists, 47–48
types, 249–250
using returned, 169–170
Data source names (DSNs), creating, 175–176
Database-driven WAP, 159–170
ADO (ActiveX Data Objects), 168–170
ASP (active server pages), 161–167
Databases
building, 173–176
querying, 169
with ShortDescription fields added, 202–203
Deck
footers, 73
headers, 72
Decks, 13
inside first, 75
using multiple, 73–74
Declarations
function, 260–262
variable, 248
XML, 71
Declaring variables, 123
Decrement operators, 126–128

Default index page, 66
Default.wml, 177–178
Design mistakes, common, 37
Design process, application, 32–37
    customers for, 32
    designing phones, 33
    drawing screens for applications, 34
    filling in gaps, 36
    micro-browser design, 32
    problems of users, 33–34
    solving problems that are visualized, 34
    testing final products, 36–37
    testing mock-ups, 34
    testing prototypes for usability, 36
    writing prototypes, 35–36
    writing pseudo codes, 35
Designing
    good WAP applications, 30–37
    phones, 33
    for users, 25–27
Device limitations, 31
DHTML (dynamic HTML), 118
Dialogs libraries, 146
Digital certificates, 218
Directives, pragma, 154
Display cards, 49–50
Do element, 94–97
Document prologue, 70–72
Document type declaration (DTD), 71
Dollar ($) signs, 104
Dot WAP 2.0, 57
Dreamweaver, Nokia WML Studio for, 56–57
DSNs (data source names), creating, 175–176
DTD (document type declaration), 71
Dynamic HTML (DHTML), 118

E

Ease of use, 25
Editors and emulators, 55
Editors, WAP, 55–57
    Dot WAP 2.0, 57
    MobileJAG, 57
    Nokia WML Studio for Dreamweaver, 56–57
    ScriptBuilder 3.0 WML Extension, 57
    Textpad, 56
    WAPTor, 56
    WML Editor, 56
    WML Express, 55
    XML Writer, 57

ElementAt function, 277
Elements
    card, 105
    Do, 94–97
    do, 95, 96
    function, 277–278
    optgroup, 88, 90
    option, 90
    placing event types directly into card, 102
    postfield, 166
    template, 92–97
Employees, wireless, 19–20
Empty statements, 262
Emptyok, 113
Emulators
    editors and, 55
    Ericsson R380, 58
Emulators, WAP, 58–60
    Ericsson R380 Emulator, 58
    EzWAP 1.0, 58
    M3Gate, 60
    Opera, 59
    Pyweb Deck-it, 60
    WAPalizer, 58
    WAPEmulator, 60
    WAPMan for Windows 95/98/NT, 59
    WAPsilon, 59
    WAPsody, 59
    WinWAP, 59
    Wireless Companion, 60
    Yospace, 58
Encryption, brief history of, 216–218
End of File (EOF), 180
Entries
    data field, 45
    password text, 44
    text, 43–45
Entry cards, 49
EOF (End of File), 180
Ericsson Developer Zone, 58
Ericsson R380 Emulator, 58
Errors, continue statements and generating, 265
EscapeString function, 285
Event handler defined, 98
Event types, placing directly into card elements, 102
Events, 98–104
    onenterbackward, 98–101
    onenterforward, 101–102
    onpick attribute, 102–103
    timer, 103
Example, Hello World, 70–73

Index

Example, services site, 73–84
    building services sites, 74–75
    contents of cards, 75–76
    creating links, 80–81
    graphics, 76–78
    inside first deck, 75
    services site with graphics, 78–80
    templates, 83–84
    using multiple decks, 73–74
    WML site with links, 81–83
Exit function, 268
Expression statements, 262
Expressions, 134, 259
External files, 155–156, 250–251
External function calls, 261
EzWAP 1.0, 58

F

Features, solutions versus, 26–27
Field entry, data, 45
Field, password data entry, 178
Files, external, 155–156, 250–251
Final products, testing, 36–37
Find function, 278
First card, 72
First deck, inside, 75
Float function, 268
Float libraries, 146–147
Float library, 272–275
Floating-point constants, 244–245
Floor function, 273
Footer, deck, 73
For statements, 137–138, 264
Format attributes, 116
Format codes, WML, 115
Format function, 278–280
Formats
    WAP (Wireless Application Protocol), 66
    WBMP (wireless bitmap image), 62
Formatting, data, 112–116
Forms and user input, interactivity, 85–116
Forum, WAP, 7
Function calls, 260–261
    external, 261
    library, 262
    local script, 261
Function declarations, 260–262
Functions, 141–145, 260–262
    arguments, 141
    calling, 142–145

    and notational conventions, 266
    parameters, 141–142
    stub, 35
Future of WAP (Wireless Application Protocol), 233–240
    Bluetooth technology, 235–237
    bringing it all together, 239–240
    new slant on walkie/talkies, 237–238
    technology with users in mind, 234–235
    Telematics, 238–239
    VoiceXML, 237–238

G

General Packet Radio Service (GPRS), 11, 21
Generating errors, 265
GetBase function, 285
GetCurrentCard function, 291
GetFragment function, 285–286
GetHost function, 286
GetParameters function, 286
GetPath function, 287
GetPort function, 287
GetQuery function, 287–288
GetReferer function, 288
GetScheme function, 288
GetVar function, 291
Global Positioning System (GPS), 6, 18, 228, 230, 231
Global System for Mobile Communications (GSM), 6
Go function, 291–292
GPRS (General Packet Radio Service), 11, 21
GPS (Global Positioning System), 6, 18, 228, 230, 231
Graphics, 51, 76–78
    <img> tag, 77–78
    services site with, 78–80
    using, 11–12
Groups, option, 88–92
GSM (Global System for Mobile Communications), 6

H

Handheld Device Markup Language (HDML), 6, 7, 27
Handheld Device Transport Protocol (HDTP), 6
Handler, event, 98
Handshakes, 220–221

HDML (Handheld Device Markup Language), 6, 7, 27
HDTP (Handheld Device Transport Protocol), 6
Headers, deck, 72
Hello World example, 70–73
    deck footer, 73
    deck header, 72
    document prologue, 70–72
    first card, 72
    second card, 72–73
Help card, 76
Help type, 95–96
High latency, 12–13
HTML conversion, demonstration, 202–210
HTML (Hypertext Markup Language), 3
HTML to WML conversion rules, 199–202
HTML Web sites, converting existing to WAP, 196–202
HTTP (Hypertext Transfer Protocol), 3
Hypertext Markup Language (HTML), 3

 I

If statements, 135–137, 263–264
IIS (Internet Information Server), 162
Images
    branding, 77
    converting, 62
    corporate, 77
<img> tag, 77–78
Increment operators, 126–128
Index page, default, 66
Information, location-sensitive, 228–229
Input facilities, limited, 14
Input tags, 110–116
InsertAt function, 280
Int function, 273
Integer constants, 244
Integrated development environments (IDES), 60–61
Intelligent Terminal Transfer Protocol (ITTP), 6
Interactivity, forms and user input, 85–116
Interfaces, user, 39–51
    Back button, 50–51
    graphics, 51
    low bandwidth, 42
    small screen size, 42–43
    text entry, 43–45
    types of WML cards, 47–50
    using caches, 45–46
Interfaces, writing generic WML, 28–29

Internet Information Server (IIS), 162
Internet; *See* Web
Invalid variables, 246
IsEmpty function, 280–281
IsFloat function, 268
IsInt function, 269
Isvalid
    functions, 288–289
    operators, 133, 259
ITTP (Intelligent Terminal Transfer Protocol), 6

 J

JavaScript principles, 163

 K

Keystrokes, number of, 43
Killer applications, 30

 L

Lang library, 147–148, 267–272
Language, WMLScript is client-side scripting, 118
Latencies
    average, 12
    high, 12–13
Length function, 281
Libraries, 266–295
    Console, 295
    Dialogs, 146, 293–294
    Float, 146–147, 272–275
    Lang, 147–148, 267–272
    notational conventions, 266
    standard, 145–152
    String, 149–150, 275–284
    URL, 150–151, 284–290
    WMLBrowser, 152, 290–293
Library function calls, 262
Library functions, Console
    print, 295
    printLn, 295
Library functions, Dialogs
    alert, 293–294
    confirm, 294
    prompt, 294
Library functions, Float
    ceil, 273

floor, 273
int, 273
maxFloat, 274
minFloat, 274
pow, 274
round, 274–275
sqrt, 275
Library functions, Lang
    abort, 267
    abs, 267
    characterSet, 267
    exit, 268
    float, 268
    isFloat, 268
    isInt, 269
    max, 269
    maxInt, 269–270
    min, 270
    minInt, 270
    parseFloat, 270–271
    parseInt, 271
    random, 272
    seed, 272
Library functions, String
    charAt, 276
    compare, 276
    elementAt, 277
    elements, 277–278
    find, 278
    format, 278–280
    insertAt, 280
    isEmpty, 280–281
    length, 281
    removeAt, 281–282
    replace, 282
    replaceAt, 282–283
    squeeze, 283
    subString, 283–284
    toString, 284
    trim, 284
Library functions, URL
    escapeString, 285
    getBase, 285
    getFragment, 285–286
    getHost, 286
    getParameters, 286
    getPath, 287
    getPort, 287
    getQuery, 287–288
    getReferer, 288
    getScheme, 288
    isValid, 288–289
    loadString, 289–290
    resolve, 290
    unescapeString, 290
Library functions, WMLBrowser
    getCurrentCard, 291
    getVar, 291
    go, 291–292
    newContext, 292
    prev, 292
    refresh, 292–293
    setVar, 293
Lifetime, variable, 248
Limitations
    content, 31
    device, 31
Line breaks, 120
    whitespace and, 242
Line comments, 243
Links
    creating, 80–81
    WML site with, 81–83
Lists, data, 47–48
Literals, 244
LoadString function, 289–290
Local script function calls, 261
Location-based services, 18
Location-sensitive information, 228–229
Logical operators, 128–129, 255–256
Login.asp, 178–181
Loop statements, skipping unnecessary, 138–139
Loops, stopping, 138–139
Low bandwidth, 11–12, 42

M

M-commerce and security, 213–221
M3Gate, 60
Market micro-browsers, targeting, 29–30
Max function, 269
MaxFloat function, 274
MaxInt function, 269–270
Memory, limited, 14
Menu, options, 86–92
Menu, select, 102
Menu.asp, 181–184
Message response, bad, 227
Messages, push, 225–228
Metadata, 157, 252

Metapragmas, 252
    content, 252
    file's scheme, 252
    property name, 252
Micro-browser
    design, 32
    issues today, 27–30
    WAP, 27
Micro-browsers, targeting market, 29–30
Min function, 270
MinFloat function, 274
MinInt function, 270
Mistakes, common design, 37
Mobile commerce, 2
Mobile Station International Subscriber Directory Number (MSISDN), 225
MobileJAG, 57
Mock-ups, testing, 34
Model, ASP object, 164–167
    Application objects, 167
    Request object, 165–166
    Response object, 164–165
    Server objects, 167
    Session object, 166
Moore's Law, 218
MSISDN (Mobile Station International Subscriber Directory Number), 225
Multiple decks, using, 73–74

N

Name attributes, 111
Names, punctuation around variable, 107
Navigation-Only card, 48
Networks, restrictions of wireless, 10–11
Newcontext, 105
NewContext function, 292
Nokia
    SDK version 2.0, 94
    selection on, 87
    WML Studio for Dreamweaver, 56–57
Noop statement, 95
Notational conventions, 266

O

Object model, ASP, 164–167
Objects
    Application, 167
    Recordset, 168, 169, 190
    Request, 165–166
    Response, 164–165
    Server, 167
    Session, 166
Onenterbackward events, 98–101
Onenterforward, 107
Onenterforward events, 101–102
Onpick attribute, 90
Onpick attributes, 92, 102–103
Ontimers, 103–104
Opera, 59
Operator precedence, 134
Operators, 124–135, 253–259
    arithmetic, 125, 254–255
    assignment, 125, 253–254
    bitwise, 126
    comma, 257
    commas, 131–132
    comparison, 128, 129–131, 256–257
    conditional, 133–134, 258
    decrement, 126–128
    increment, 126–128
    isvalid, 133, 259
    logical, 128–129, 255–256
    string, 256
    string concatenation, 131
    typeof, 132–133, 258
Optgroup
    elements, 88, 90
    statements, 92
Option
    elements, 90
    groups, 88–92
    statements, 92
Options
    Change, 49
    Pick, 48
Options menu (select), 86–92
Options statement, 86
Options type, 96–97
OTA (Over The Air) protocol, 225

P

Pages
    are called cards in WAP, 69
    default index, 66
    sub, 67
PAP (Push Access Protocol), 225
ParseFloat function, 270–271
ParseInt function, 271

Password
　　data entry fields, 178
　　text entries, 44
　　types, 178
Passwords, 99
PDAs (personal digital assistants), 2, 3, 5
Personal Web Server (PWS), 162
Personalization, application, 44
Phone.com, selection on, 88
Phones, designing, 33
PI (Push Initiators), 225
Pick option, 48
Point of view, user's, 24–27
Postfield elements, 166
Postfix defined, 127
Pow function, 274
PPG (Push Proxy Gateways), 225
Pragma directives, 154
Pragmas, 154–157, 250–252
　　access control, 156–157, 251–252
　　external files, 155–156, 250–251
　　metadata, 157, 252
Prefix defined, 127
Prefs.asp, 184–186
Prev function, 292
Principles
　　general coding, 157–158
　　JavaScript, 163
　　VBScript, 163
print function, 295
PrintLn function, 295
Privacy, user, 231
Problems, solving, 34
Products, testing final, 36–37
Prologues, document, 70–72
Prompt function, 294
Properties.asp, 202
Protocols
　　OTA (Over The Air), 225
　　Push Access, 225
Prototypes
　　testing for usability, 36
　　writing, 35–36
Pseudo codes, writing, 35
Public key cryptography, 217–218
Punctuation around variable names, 107
Push
　　cancellation, 226
　　framework, 224–228
　　submission, 225
　　and telematics together, 230–231
Push Access Protocol (PAP), 225

Push Initiators (PI), 225
Push messages, 225–228
　　bad-message response, 227
　　client-capability query, 227
　　push cancellation, 226
　　push submission, 225
　　result-notification, 225–226
　　status query, 226
Push Proxy Gateways (PPG), 225
Push technology, 224–228
　　and telematics, 223
PWS (Personal Web Server), 162
Pyweb Deck-it, 60

Q

Queries
　　client-capability, 227
　　status, 226
Querying databases, 169

R

Random function, 272
Recordset object, 168, 169, 190
Reference, WMLScript, 241–295
Referring to variables, $ signs and, 104
Refresh function, 292–293
Refresh tag, 106
RemoveAt function, 281–282
Replace function, 282
ReplaceAt function, 282–283
Request object, 165–166
Reserved words, 140–141, 247
Reset type, 96
Resolve function, 290
Response object, 164–165
Results.asp, 190–193
Return statements, 266
Round function, 274–275

S

Scope, variable, 248
Screen size, small, 42–43
Screens
　　drawing for applications, 34
　　splash, 77, 103
ScriptBuilder 3.0 WML Extension, 57

Scripting language, WMLScript is
    client-side, 118
Second card, 72–73
Secure Socket Layer (SSL), 219
Security
    acceptable levels of, 215
    and M-commerce, 213–221
    reasons for having, 214–216
    types of, 214–216
    of WAP, 215–216
Seed function, 272
Select
    menu, 102
    statement, 88
Server objects, 167
Services
    location-based, 18
    WAP, 15–20
Services site example, 73–84
    building services sites, 74–75
    contents of cards, 75–76
    creating links, 80–81
    graphics, 76–78
    inside first deck, 75
    services site with graphics, 78–80
    templates, 83–84
    using multiple decks, 73–74
    WML site with links, 81–83
Services sites
    building, 74–75
    with graphics, 78–80
Session object, 166
Setting variables, miscellaneous ways of,
    108–110
SetVar function, 293
Setvar tag, 106
Short Message Services (SMS), 4, 7, 11, 219
$ signs, 104
Signs, $, 104
Sites
    building services, 74–75
    converting existing Web, 195–211
    services, 78–80
    web, 118
    WML, 81–83
Skipping unnecessary loop statements, 138–139
SMS (Short Message Services), 4, 7, 11, 219
Software developer kits (SDKS), 60–61
Software, WAP development tools and, 53–63
Solutions versus features, 26–27
Specification, WML is strict XML, 199
Splash screens, 77, 103

Sqrt function, 275
Squeeze function, 283
SSL (Secure Socket Layer), 219
Standard libraries, 145–152
Statements, 122, 262–266
    for, 137–138, 264
    block, 262–263
    break, 138–139, 265
    continue, 139, 265
    empty, 262
    expression, 262
    if, 135–137, 263–264
    noop, 95
    optgroup, 92
    options, 86, 92
    return, 266
    select, 88
    skipping unnecessary loop, 138–139
    template, 94
    variable, 263
    while, 137, 264
Static
    cards, 67
    data hard-coded in cards, 160
Status query, 226
Stopping loops, 138–139
String
    concatenation, 131
    constants, 245–246
    operators, 256
String library, 149–150, 275–284
Stub function defined, 35
Sub-pages, 67
SubString function, 283–284
Symmetric ciphers, 216, 217
Symmetric cryptography, 217

▼ T

Tagged Text Markup Language (TTML), 7
Tags
    <img>, 77–78
    input, 110–116
    refresh, 106
    setvar, 106
    template, 93
Targeting market micro-browsers, 29–30
TCP (Transmission Control Protocol), 8
Technology
    Bluetooth, 234, 235–237
    push, 223, 224–228

Index

with users in mind, 234–235
Telematics, 228–230, 238–239
    applications for, 229–230
    defined, 228
    and push technology, 223
    and push together, 230–231
Template
    codes, 92, 93
    elements, 92–97
    statements, 94
Template tags, 93
Templates, 83–84
Testing
    final products, 36–37
    mock-ups, 34
    prototypes for usability, 36
Text entry, 43–45
    application personalization, 44
    data field entry, 45
    number of keystrokes, 43
    password, 44
    password text entry, 44
Text type, 178
Textpad, 56
Timer events, 103
Tools and software, WAP development, 53–63
ToString function, 284
Transmission Control Protocol (TCP), 8
Trim function, 284
TTML (Tagged Text Markup Language), 7
Typeof operators, 132–133, 258

U

UnescapeString function, 290
URL library, 150–151, 284–290
Use, ease of, 25
User agents defined, 2
User input, forms and, 85–116
User interface, 39–51
    Back button, 50–51
    basics, 40–42
    graphics, 51
    low bandwidth, 42
    small screen size, 42–43
    text entry, 43–45
    types of WML cards, 47–50
    using caches, 45–46
User privacy, 231
Users
    point of view, 24–27
    problems of, 33–34
    and technology, 234–235
Users, designing for, 25–27
    critical, 26
    fun, 26
    solutions versus features, 26–27
    useful, 26

V

Value attributes, 111
Variable declarations, 248
Variable lifetime, 248
Variable names, punctuation around, 107
Variable scope, 124, 248
Variable statements, 263
Variables, 104–110, 123–124, 248
    Boolean, 246
    declaring, 123
    invalid, 246
    miscellaneous ways of setting, 108–110
    $ signs and referring to, 104
    used for first time, 123
    using, 105–108
VBScript principles, 163
VoiceXML, 237–238

W

WAE (Wireless Application Environment), 8, 9
Walkie/talkies, new slant on, 237–238
WAP applications, 17
    designing good, 30–37
    dynamic, 171–193
WAP bitmap (WBMP), 77
WAP, database-driven, 159–170
    ADO (ActiveX Data Objects), 168–170
    ASP (active server pages), 161–167
WAP editors, 55–57
WAP emulators, 58–60
WAP format, 66
WAP Forum, 7
WAP micro-browser issues today, 27–30
WAP services, 15–20
    customer care, 18–19
    location-based services, 18
    wireless employees, 19–20
WAP (Wireless Application Protocol), 2
    architecture, 8–15
    and ASP, 162–164

benefits for consumers, 5–6
business case for, 15–21
converting existing Web sites to, 196–202
defined, 2–6
development tools and software, 53–63
evolution of, 10
future of, 21–22, 233–240
goals of, 7
history of, 6–8
idea of, 7–8
importance of, 3–5
introducing, 1–22
making of good applications, 23–37
model, 9
pages are called cards in, 69
security of, 215–216
software, 53–63
time before, 6–7
and Web, 67–68
why have, 20–21
WAP, World-wide-Dance-Web for, 172–193
    building databases, 173–176
    creating DSNs (data source names), 175–176
    data flow, 172–173
    writing code, 176–193
WAPalizer, 58
WAPEmulator, 60
WAPMan for Windows 95/98/NT, 59
WAPsilon, 59
WAPsody, 59
WAPTor, 56
WBMP types, specification of well-defined, 62
WBMP (WAP bitmap), 77
WDP (Wireless Datagram Protocol), 9
Web sites, 118
    converting existing, 195–211
Web sites to WAP, converting existing to HTML, 196–202
    configurable converters, 197–198
    do-it-yourself, 198–202
    fully automated converters, 197
    ideal sites for conversion, 196–197
    methods of conversion, 197–202
Web, WAP and, 67–68
While statements, 137, 264
Whitespace, 120
Whitespace and line breaks, 242
Windows 95/98/NT, WAPMan for, 59
WinWAP, 59
Wireless Application Environment (WAE), 8, 9

Wireless Application Protocol (WAP); *See* WAP (Wireless Application Protocol)
Wireless bitmap image (WBMP) format, 62
Wireless Companion, 60
Wireless Datagram Protocol (WDP), 9
Wireless employees, 19–20
Wireless Markup Language (WML); *See* WML (Wireless Markup Language)
Wireless networks, restrictions of, 10–11
    high latency, 12–13
    less connection stability, 13
    limited battery power, 14–15
    limited CPU, 14
    limited input facilities, 14
    limited memory, 14
    low bandwidth, 11–12
    small display, 13–14
    unpredictable bearer availability, 13
Wireless Session Protocol (WSP), 8, 11, 157
Wireless Telephony Application Interface (WTAI), 8, 12–13
Wireless Transaction Protocol (WTP), 8, 11, 13
Wireless Transport Layer Security (WTLS), 9, 215, 219–221
WML cards, types of, 47–50
    choice cards, 47–49
    display cards, 49–50
    entry cards, 49
WML code, writing, 68–69
WML conversion rules, HTML to, 199–202
WML Editor, 56
WML Express, 55
WML interfaces, writing generic, 28–29
WML Studio for Dreamweaver, Nokia, 56–57
WML (Wireless Markup Language), 8, 9, 10
    basics, 66–69
    format codes, 115
    is strict XML specification, 199
    site with links, 81–83
WML (Wireless Markup Language), working with, 65–84
    Hello World example, 70–73
    services site example, 73–84
WMLBrowser libraries, 152, 290–293
WMLScript, 8, 10, 11
    client-side scripting language, 118
    compiler, 242
    defined, 118–119
WMLScript, adding functionality with, 117–158
    arrays, 152–154
    control constructs, 135–139
    functions, 141–145

Index

general coding principles, 157–158
operators, 124–135
pragmas, 154–157
reserved words, 140–141
rules of WMLScript, 119–122
standard libraries, 145–152
variables, 123–124
WMLScript defined, 118–119
WMLScript reference, 241–295
case sensitivity, 242
comments, 243
constants, 244–246
data types, 249–250
expressions, 259
functions, 260–262
integer constants, 244
libraries, 266–295
operators, 253–259
pragmas, 250–252
reserved words, 247
statements, 262–266
variables, 248
whitespace and line breaks, 242
WMLScript, rules of, 119–122
case sensitivity, 120
code blocks, 122
comments, 120–121
line breaks, 120
statements, 122
whitespace, 120
Words, reserved, 140–141, 247
World-wide-Dance-Web for WAP, 172–193
World-wide-Dance-Web.com, 172
World Wide Web; *See* Web

Writing
generic WML interfaces, 28–29
prototypes, 35–36
pseudo codes, 35
WML code, 68–69
Writing code, 176–193
chprefs.asp, 186–188
Default.wml, 177–178
login.asp, 178–181
menu.asp, 181–184
prefs.asp, 184–186
results.asp, 190–193
wrprefs.asp, 188–190
Wrprefs.asp, 188–190
WSP (Wireless Session Protocol), 8, 11, 157
WTAI (Wireless Telephony Application Interface), 8, 12–13
WTLS (Wireless Transport Layer Security), 9, 215, 219–221
WTP (Wireless Transaction Protocol), 8, 11, 13

X

X, referring to letter, 104
XML declaration, 71
XML specification, WML is strict, 199
XML Writer, 57

Y

Yospace, 58

INTERNATIONAL CONTACT INFORMATION

AUSTRALIA
McGraw-Hill Book Company Australia Pty. Ltd.
TEL +61-2-9417-9899
FAX +61-2-9417-5687
http://www.mcgraw-hill.com.au
books-it_sydney@mcgraw-hill.com

CANADA
McGraw-Hill Ryerson Ltd.
TEL +905-430-5000
FAX +905-430-5020
http://www.mcgrawhill.ca

GREECE, MIDDLE EAST, NORTHERN AFRICA
McGraw-Hill Hellas
TEL +30-1-656-0990-3-4
FAX +30-1-654-5525

MEXICO (Also serving Latin America)
McGraw-Hill Interamericana Editores S.A. de C.V.
TEL +525-117-1583
FAX +525-117-1589
http://www.mcgraw-hill.com.mx
fernando_castellanos@mcgraw-hill.com

SINGAPORE (Serving Asia)
McGraw-Hill Book Company
TEL +65-863-1580
FAX +65-862-3354
http://www.mcgraw-hill.com.sg
mghasia@mcgraw-hill.com

SOUTH AFRICA
McGraw-Hill South Africa
TEL +27-11-622-7512
FAX +27-11-622-9045
robyn_swanepoel@mcgraw-hill.com

UNITED KINGDOM & EUROPE (Excluding Southern Europe)
McGraw-Hill Education Europe
TEL +44-1-628-502500
FAX +44-1-628-770224
http://www.mcgraw-hill.co.uk
computing_neurope@mcgraw-hill.com

ALL OTHER INQUIRIES Contact:
Osborne/McGraw-Hill
TEL +1-510-549-6600
FAX +1-510-883-7600
http://www.osborne.com
omg_international@mcgraw-hill.com